ESSENTIAL LABORATORY MANUAL FOR GENERAL, ORGANIC, AND BIOLOGICAL CHEMISTRY

SECOND EDITION

KAREN C. TIMBERLAKE
Los Angeles Valley College

WITH SIGNIFICANT CONTRIBUTION FROM
GARY BURCE,
Ohlone College

Prentice Hall

Boston Columbus Indianapolis New York San Francisco Upper Saddle River
Amsterdam Cape Town Dubai London Madrid Milan Munich Paris Montréal Toronto
Delhi Mexico City São Paulo Sydney Hong Kong Seoul Singapore Taipei Tokyo

Editor in Chief, Chemistry and Geosciences: Nicole Folchetti
Marketing Manager: Erin Gardner
Assistant Editor: Jessica Neumann
Managing Editor, Chemistry and Geosciences: Gina M. Cheselka
Project Manager: Traci Douglas
Operations Specialist: Maura Zaldivar
Supplement Cover Designer: Tom Nery
Cover Photo Credit: Shutterstock Stock Photo

© 2011, 2007
Pearson Prentice Hall
Pearson Education, Inc.
Upper Saddle River, New Jersey 07458

All rights reserved. No part of this book may be reproduced, in any form or by any means, without permission in writing from the publisher.

Pearson Prentice Hall™ is a trademark of Pearson Education, Inc.

The author and publisher of this book have used their best efforts in preparing this book. These efforts include the development, research, and testing of the theories and programs to determine their effectiveness. The author and publisher make no warranty of any kind, expressed or implied, with regard to these programs or the documentation contained in this book. The author and publisher shall not be liable in any event for incidental or consequential damages in connection with, or arising out of, the furnishing, performance, or use of these programs.

This work is protected by United States copyright laws and is provided solely for teaching courses and assessing student learning. Dissemination or sale of any part of this work (including on the World Wide Web) will destroy the integrity of the work and is not permitted. The work and materials from it should never be made available except by instructors using the accompanying text in their classes. All recipients of this work are expected to abide by these restrictions and to honor the intended pedagogical purposes and the needs of other instructors who rely on these materials.

Printed in the United States of America

ISBN-13: 978-0-13-605547-1
ISBN-10: 0-13-605547-8

9 2021

Preface

This *Essential Laboratory Manual for General, Organic, and Biological Chemistry* consolidates in one place the experiments that I have found to be most valuable to my students. It provides experiments that illustrate each of the chemical principles we discuss from the first day of class. I have also taken care to make each experiment workable as well as providing critical thinking for the student. The goals of this laboratory manual address the following areas:

1. **Experiments relate to basic concepts of chemistry and health.** Experiments are designed to illustrate the chemical principles we discuss in our classes. They include experiments that relate to health and medicine, and often use common materials that are familiar to students.

2. **Experiments are flexible.** Each experiment includes a flexible group of sections, which allows instructors to select the sections to fit into their weekly laboratory schedule. Lab times and comments are given for each.

3. **Safety.** A detailed safety section in the preface includes a safety quiz. The aim here is to highlight the safety and equipment preparation on the first day of lab. In addition, each lab contains reminders of safety behavior. Students are reminded to wear goggles for every lab session. Some experiments are recommended as instructor demonstrations.

4. **Experiment format provides clear instructions and evaluation.** Each lab begins with a set of goals, a discussion of the topics, and examples of calculations. The report pages begin with pre-lab questions to prepare students for lab work. Students obtain data, draw graphs, make calculations, and write conclusions about their results. Each lab contains questions and problems that require the student to discuss the experiment, make additional calculations, and use critical thinking to apply concepts to real life.

5. **Stockroom preparation of chemicals.** Materials for each experiment are listed in the appendix with amounts given for 20 students working in pairs. Most lab sessions use standard lab equipment and chemicals that are readily available and inexpensive. In some cases students bring samples from home.

I hope that this laboratory manual will help you in your chemistry instruction and that students will find they learn chemistry by participating in the laboratory experience.

Karen C. Timberlake
Los Angeles Valley College
Valley Glen, CA 91401

To the Student

Here you are in a chemistry laboratory with your laboratory book in front of you. Perhaps you have already been assigned a laboratory drawer, full of glassware and equipment you may never have seen before. Looking around the laboratory, you may see bottles of chemical compounds, balances, burners, and other equipment that you are going to use. This may very well be your first experience with experimental procedures. At this point you may have some questions about what is expected of you. This laboratory manual is written with those considerations in mind.

The activities in this manual were written specifically to parallel the chemistry you are learning in the lecture portion of class. The manual sections and pages that are directly relevant to each experiment are identified at the beginning of the experiment. Many of the laboratory activities include materials that may be familiar to you, such as household products, diet drinks, cabbage juice, antacids, and aspirin. In this way, chemical topics are related to the real world and to your own science experience. Some of the labs teach basic skills; others encourage you to extend your scientific curiosity beyond the lab.

It is important to realize that the value of the laboratory experience depends on the time and effort you invest in it. Only then will you find that the laboratory can be a valuable learning experience and an integral part of the chemistry class. The laboratory gives you an opportunity to go beyond the lectures and words in your textbook and experience the scientific process from which conclusions and theories concerning chemical behavior are drawn. In some experiments, the concepts are correlated with health and biological concepts. Chemistry is not an inanimate science, but one that helps us to understand the behavior of living systems.

Using This Laboratory Manual

Each experiment begins with learning goals to give you an overview of the topics you will be studying in that experiment. Each experiment is correlated to concepts you are currently learning in your chemistry class. Your instructor will indicate which activities you are to do. At the beginning of each experiment, you will also find a list of the materials needed for each activity.

The experimental procedures are written to guide you through each laboratory activity. They are written in a tabular modified SOP (Standard Operating Procedure) format. Each line of an SOP table is a single action. Complete the actions in the order in the table. When you are finished with the action, check the completed box. In this way, you will know exactly which steps you have completed.

Section	Step	Procedure	Check when finished
A.4	1	After the water in the water bath has boiled for 10 minutes or more, use a thermometer to measure the temperature of the boiling water.	X
A.4	2	Record. (This is also the initial temperature of your metal object.)	

Example of tabular SOP with the first step checked.

If you will need to perform the same steps several times (e. g. the same steps for several different samples), the check box will be divided into the appropriate number of smaller boxes. Each smaller box should be checked as it is completed.

| A.1 | 6 | After 5 minutes, use clean droppers to transfer four drops of each mixture to a spot plate (or plastic sheet). | X | |
| A.1 | 7 | Add 1 drop of iodine solution to each. A blue-black color | X | |

		indicates that starch remains in the mixture.		
A.1	8	Record your observations	X	
A.1	9	Repeat steps 6 through 8 at 10 minutes.	X	

In this Example of a tabular SOP, steps 6 through 8 must be performed at both 5 minutes and 10 minutes. They have already been performed at 5 minutes and need to be performed at 10 minutes.

When you are ready to begin the lab, remove the report sheet at the end of the laboratory instructions. Place it next to the procedures for that section. Read and measure carefully, report your data, and follow instructions to complete the necessary calculations. You may also be asked to answer some or all of the follow-up questions and problems designed to test your understanding of the concepts from the experiment.

It is my hope that the laboratory experience will help illuminate the concepts you are learning in the classroom. The experimental process can help make chemistry a real and exciting part of your life and provide you with skills necessary for your future.

Pre-Lab Study Questions

There are Pre-Lab Study questions on a separate page at the beginning of each laboratory report section. These questions should be completed before you come to lab. Your laboratory instructor may require that you hand in the study questions before you begin your laboratory work. If so, it is easy to remove the Pre-Lab section from the lab manual.

Contents

Working Safely in the Laboratory — ix
Commitment to Safety in the Laboratory — xiv
A Visual Guide to Laboratory Equipment — xvi
Graphing Experimental Data — xviii
Using the Laboratory Burner — xx

PART I: DRY LABS

D-1 Conversion Factors in Calculations D-1
  A. Rounding Off D-1
  B. Significant Figures in Calculations D-2
  C. Conversion Factors for Length D-3
  D. Conversion Factors for Volume D-3
  E. Conversion Factors for Mass D-4
  F. Converting Temperature D-4

D-2 Compounds and Their Formulas D-19
  A. Electron-Dot Structures D-20
  B. Ionic Compounds and Formulas D-21
  C. Ionic Compounds with Transition Metals D-22
  D. Ionic Compounds with Polyatomic Ions D-22
  E. Covalent (Molecular) Compounds D-23

D-3 Properties and Structures of Alkanes D-35
  A. Color, Odor, and Physical State D-35
  B. Solubility D-36
  C. Combustion D-36
  D. Structures of Alkanes D-36
  E. Constitutional Isomers D-37
  F. Cycloalkanes D-38

PART II: WET LABS

1 Measurement and Significant Figures 1
  A. Measuring Length 3
  B. Measuring Volume 4
  C. Measuring Mass 5

2 Density and Specific Gravity 13
  A. Density of a Solid 13
  B. Density of a Liquid 13
  C. Specific Gravity 14

3 Atomic Structure and Electron Arrangement 25
  A. Physical Properties of Elements 25
  B. Periodic Table 26
  C. Subatomic Particles 26
  D. Isotopes 26
  E. Flame Tests 27
  F. Drawing Models of Atoms 27

4 Nuclear Radiation 41
  A. Background Count 42
  B. Radiation from Radioactive Sources 43
  C. Effect of Shielding, Time, and Distance 43

5	**Chemical Reactions and Equations**	53
	A. Magnesium and Oxygen 55	
	B. Zinc and Copper (II) Sulfate 56	
	C. Metals and HCl 57	
	D. Reactions of Ionic Compounds 57	
	E. Sodium Carbonate and HCl 59	
6	**Moles and Chemical Formulas**	67
	A. Finding the Simplest Formula 67	
	B. Formula of a Hydrate 68	
7	**Energy and Matter**	83
	A. A Heating Curve for Water 83	
	B. Energy in Changes of State 84	
	C. Food Calories 85	
	D. Exothermic and Endothermic Reactions 85	
8	**Gas Laws: Boyle's and Charles'**	99
	A. Boyle's Law 100	
	B. Charles' Law 100	
9	**Partial Pressures of Oxygen, Nitrogen, and Carbon Dioxide**	117
	A. Partial Pressures of Oxygen and Nitrogen in Air 117	
	B, C. Carbon Dioxide in the Atmosphere and Expired Air 118	
10	**Solutions**	129
	A. Polarity of Solutes and Solvents 129	
	B. Solubility of KNO_3 130	
	C. Concentration of a Sodium Chloride Solution 130	
11	**Testing for Cations and Anions**	143
	A. Tests for Positive Ions (Cations) 143	
	B. Tests for Negative Ions (Anions) 144	
	C. Writing the Formulas of Your Unknown Salt 144	
12	**Solutions, Colloids, and Suspensions**	155
	A. Identification Tests 155	
	B. Osmosis and Dialysis 155	
	C. Filtration 156	
13	**Acids and Bases**	169
	A. pH Color Using Red Cabbage Indicator 170	
	B. Measuring pH 170	
	C. Acetic Acid in Vinegar 171	
14	**Alcohols, Aldehydes, and Ketones**	185
	A. Structures of Alcohols and Phenol 185	
	B. Properties of Alcohols and Phenol 186	
	C. Oxidation of Alcohols 186	
	D. Properties of Aldehydes and Ketones 187	
	E. Iodoform Test for Methyl Ketones 187	
	F. Oxidation of Aldehydes and Ketones 188	
15	**Carboxylic Acids and Esters**	201
	A. Carboxylic Acids and Their Salts 202	
	B. Esters 202	
	C. Basic Hydrolysis of Esters 203	

16 Carbohydrates — 217
A. Monosaccharides 217
B. Disaccharides 219
C. Polysaccharides 219
D. Benedict's Test for Reducing Sugars 221
E. Seliwanoff's Test for Ketoses 222
F. Fermentation Test 222
G. Iodine Test for Polysaccharides 222
H. Hydrolysis of Disaccharides and Polysaccharides 222

17 Lipids — 239
A. Triacylglycerols 239
B. Physical Properties of Lipids and Fatty Acids 240
C. Bromine Test for Unsaturation 241
D. Preparation of Hand Lotion 242

18 Saponification and Soaps — 255
A. Saponification 255
A. Saponification: Preparation of Soap 255
B. Properties of Soap and Detergents 256

19 Amines and Amino Acids — 267
B. Solubility of Amines in Water 268
C. Neutralization of Amines with Acids 268
D. Amino Acids 268
D. Chromatography of Amino Acids 270

20 Peptides and Proteins — 283
A. Peptide Bonds 283
B. Structure of Proteins 284
C. Denaturation of Proteins 284
D. Isolation of Casein (Milk Protein) 284
E. Color Tests for Proteins 284

21 Enzymes — 297
A. Effect of Enzyme Concentration 298
B. Effect of Temperature 298
C. Effect of pH 299
D. Inhibition of Enzyme Activity 299

22 DNA Components and Extraction — 313
A. Components of DNA 313
B. Extraction of DNA 314

Appendix: Materials and Solutions — 325
Standard Laboratory Materials 325
Additional Materials Needed for Individual Experiments 325
Preparation of Solutions Used in the Laboratory 333

Working Safely in the Laboratory

The chemistry laboratory with its equipment, glassware, and chemicals has the potential for accidents. In order to avoid accidents, precautions must be taken by every student to ensure the safety of everyone working in the laboratory. By following the rules for handling chemicals safely and carrying out only the approved procedures, you will create a safe environment in the laboratory. After you have read the following sections, complete the safety quiz and the questions on laboratory equipment. Then sign and submit the commitment to lab safety.

A. Preparing for Laboratory Work

Pre-read Before you come to the laboratory, read the discussion of and directions for the experiment you will be doing. Make sure you know what the experiment is about before you start the actual work. If you have a question, ask your instructor to clarify the procedures.

Do assigned work only Do only the experiments that have been assigned by your instructor. No unauthorized experiments are to be carried out in the laboratory. Experiments are done at assigned times, unless you have an open lab situation. Your instructor must approve any change in procedure.

Do not work alone in a laboratory.

Safety awareness Learn the location and use of the emergency eyewash fountains, the emergency shower, fire blanket, fire extinguishers, and exits. Memorize their locations in the laboratory. Be aware of other students in the lab carrying chemicals to their desk or to a balance.

Experimental Procedures   *GOGGLES REQUIRED!*

Safety goggles must be worn all the time you are in the lab The particular type depends on state law, which usually requires industrial-quality eye protection. Contact lenses may be worn in the lab if needed for therapeutic reasons, provided that safety goggles are worn over the contact lenses. Contact lenses without goggles are dangerous because splashed chemicals make them difficult to remove. If chemicals accumulate under a lens, permanent eye damage can result. If a chemical should splash into your eyes, flood the eyes with water at the eyewash fountain. Continue to rinse with water for at least 10 minutes.

Wear protective clothing Wear sensible clothing in the laboratory. Loose sleeves, shorts, or open-toed shoes can be dangerous. A lab coat is useful in protecting clothes and covering arms. Wear shoes that cover your feet to prevent glass cuts; wear long pants and long-sleeved shirts to protect skin. Long hair should be tied back so it does not fall into chemicals or a flame from a Bunsen burner.

No food or drink is allowed at any time in the laboratory Do not let your friends or children visit while you are working in the lab; have them wait outside.

Prepare your work area Before you begin a lab, clear the lab bench or work area of all your personal items, such as backpacks, books, sweaters, and coats. Find a storage place in the lab for them. All you will need is your laboratory manual, calculator, pen or pencil, text, and equipment from your lab drawer.

B. Handling Chemicals Safely

Check labels twice Be sure you take the correct chemical. DOUBLE-CHECK THE LABEL on the bottle before you remove a chemical from its container. For example, sodium sulfate (Na_2SO_4) could be mistaken for sodium sulfite (Na_2SO_3) if the label is not read carefully.

Use small amounts of chemicals Pour or transfer a chemical into a small, clean container (beaker, test tube, flask, etc.) available in your lab drawer. To avoid contamination of the chemical reagents, never insert droppers, pipets, or spatulas into the reagent bottles. Take only the quantity of chemical you need for the experiment. Do not keep a reagent bottle at your desk; return it to its proper location in the laboratory. Label the container. Many containers have etched sections on which you can write in pencil. If not, use tape or a marking pencil.

Do not return chemicals to the original containers To avoid contamination of chemicals, dispose of used chemicals according to your instructor's instructions. Never return unused chemicals to reagent bottles. Some liquids and water-soluble compounds may be washed down the sink with plenty of water, but check with your instructor first. Dispose of organic compounds in specially marked containers in the hoods.

Do not taste chemicals; smell a chemical cautiously Never use any equipment in the drawer such as a beaker to drink from. When required to note the odor of a chemical, first take a deep breath of fresh air and hold it while you use your hand to fan some vapors toward your nose and note the odor. Do not inhale the fumes directly. If a compound gives off an irritating vapor, use it in the fume hood to avoid exposure.

Do not shake laboratory thermometers Laboratory thermometers respond quickly to the temperature of their environment. Shaking a thermometer is unnecessary and can cause breakage.

Liquid spills Spills of water or liquids at your work area or floor should be cleaned up immediately. Small spills of liquid chemicals can be cleaned up with a paper towel. Large chemical spills must be treated with absorbing material such as cat litter. Place the contaminated material in a waste disposal bag and label it. If a liquid chemical is spilled on the skin, flood immediately with water for at least 10 minutes. Any clothing soaked with a chemical must be removed immediately because an absorbed chemical can continue to damage the skin.

Mercury spills The cleanup of mercury requires special attention. Mercury spills may occur from broken thermometers. Notify your instructor immediately of any mercury spills so that special methods can be used to clean up the mercury. Place any free mercury and mercury cleanup material in special containers for mercury only.

Laboratory accidents Always notify your instructor of any chemical spill or accident in the laboratory. Broken glass can be swept up with a brush and pan and placed in a specially labeled container for broken glass. Cuts are the most common injuries in a lab. If a cut should occur, wash, elevate, and apply pressure if necessary. Always inform your instructor of any laboratory accident.

Clean up Wash glassware as you work. Begin your cleanup 15 minutes before the end of the laboratory session. Return any borrowed equipment to the stockroom. Be sure that you always turn off the gas and water at your work area. Make sure you leave a clean desk. Check the balance you used. Wash your hands before you leave the laboratory.

C. Heating Chemicals Safely

Heat only heat-resistant glassware Only glassware marked Pyrex® or Kimax® can be heated; other glassware may shatter. To heat a substance in a test tube, use a test tube holder. Holding the test tube at an angle, move it continuously through the flame. Never point the open end of the test tube at anyone or look directly into it. A hot piece of iron or glass looks the same as it does at room temperature. Place a hot object on a tile or a wire screen to cool.

Flammable liquids Never heat a flammable liquid over an open flame. If heating is necessary, your instructor will indicate the use of a steam bath or a hot plate.

Never heat a closed container When a closed system is heated, it can explode as pressure inside builds.

Fire Small fires can be extinguished by covering them with a watch glass. If a larger fire is involved, use a fire extinguisher to douse the flames. *Do not direct a fire extinguisher at other people in the laboratory.* Shut off gas burners in the laboratory. When working in a lab, tie long hair back away from the face. If someone else's clothing or hair catches on fire, get them to the floor and roll them into a fire blanket. They may also be placed under the safety shower to extinguish flames. Cold water or ice may be applied to small burns.

D. Waste Disposal

As you work in the laboratory, chemical wastes are produced. Although we will use small quantities of materials, some waste products are unavoidable. To dispose of these chemical wastes safely, you need to know some general rules for chemical waste disposal.

Metals Metals should be placed in a container to be recycled.

Nonhazardous chemical wastes Substances such as sodium chloride (NaCl) that are soluble in water and are not hazardous may be emptied into the sink. If the waste is a solid, dissolve it in water before disposal.

Hazardous chemical wastes If a substance is hazardous or not soluble in water, it must be placed in a container that is labeled for waste disposal. Your instructor will inform you if chemical wastes are hazardous and identify the proper waste containers. If you are not sure about the proper disposal of a substance, ask your instructor. The labels on a waste container should indicate if the contents are hazardous, the name of the chemical waste, and the date that the container was placed in the lab.

Hazard rating The general hazards of a chemical are presented in a spatial arrangement of numbers with the flammability rating at twelve o'clock, the reactivity rating at the three o'clock position, and the health rating at the nine o'clock position. At the six o'clock position, information may be given on the reactivity of the substance with water. If there is unusual reactivity with water, the symbol W (do not mix with water) is shown. In the laboratory, you may see these ratings in color with blue for health hazard, red for flammability, and yellow for reactivity hazards.

A chemical is assigned a relative hazard rating that ranges from 1 (little hazard) to 4 (extreme hazard). The health hazard indicates the likelihood that a material will cause injury due to exposure by contact, inhalation, or ingestion. The flammability hazard indicates the potential for burning. The reactivity hazard indicates the instability of the material by itself or with water with subsequent release of energy. Special hazards may be included such as W for reactivity with water or OX for oxidizing properties.

E. Safety Quiz

The safety quiz will review the preceding safety discussion. Circle the correct answer(s) in each of the following questions. Check your answers on page xiii.

1. Approved eye protection is to be worn
   a. For certain experiments
   b. Only for hazardous experiments
   c. All the time
2. Eating in the laboratory is
   a. Not permitted
   b. Allowed at lunch time
   c. All right if you are careful
3. If you need to smell a chemical, you should
   a. Inhale deeply over the test tube
   b. Take a breath of air and fan the vapors toward you
   c. Put some of the chemical in your hand, and smell it
4. When heating liquids in a test tube, you should
   a. Move the tube back and forth through the flame
   b. Look directly into the open end of the test tube to see what is happening
   c. Direct the open end of the tube away from other students
5. Unauthorized experiments are
   a. All right as long as they don't seem hazardous

      b. All right as long as no one finds out
      c. Not allowed

6. If a chemical is spilled on your skin, you should
      a. Wait to see if it stings
      b. Flood the area with water for 10 minutes
      c. Add another chemical to absorb it

7. When taking liquids from a reagent bottle,
      a. Insert a dropper
      b. Pour the reagent into a small container
      c. Put back what you don't use

8. In the laboratory, open-toed shoes and shorts are
      a. Okay if the weather is hot
      b. All right if you wear a lab apron
      c. Dangerous and should not be worn

9. When is it all right to taste a chemical?
      a. Never
      b. When the chemical is not hazardous
      c. When you use a clean beaker

10. After you use a reagent bottle,
      a. Keep it at your desk in case you need more
      b. Return it to its proper location
      c. Play a joke on your friends and hide it

11. Before starting an experiment,
      a. Read the entire procedure
      b. Ask your lab partner how to do the experiment
      c. Skip to the laboratory report and try to figure out what to do

12. Working alone in the laboratory without supervision is
      a. All right if the experiment is not too hazardous
      b. Not allowed
      c. Allowed if you are sure you can complete the experiment without help

13. You should wash your hands
      a. Only if they are dirty
      b. Before eating lunch in the lab
      c. Before you leave the lab

14. Personal items (books, sweater, etc.) should be
      a. Kept on your lab bench
      b. Left outside
      c. Stored out of the way, not on the lab bench

15. When you have taken too much of a chemical, you should
      a. Return the excess to the reagent bottle
      b. Store it in your lab locker for future use
      c. Discard it using proper disposal procedures

16. In the lab, you should wear
      a. Practical, protective clothing
      b. Something fashionable
      c. Shorts and loose-sleeved shirts

17. If a chemical is spilled on the table,
      a. Clean it up right away
      b. Let the stockroom help clean it up
      c. Use appropriate adsorbent if necessary

18. If mercury is spilled,
      a. Pick it up with a dropper
      b. Call your instructor
      c. Push it under the table where no one can see it

19. If your hair or shirt catches on fire, you should
      a. Use the safety shower to extinguish the flames
      b. Drop to the ground and roll
      c. Use the fire blanket to put it out

20. A hazardous waste should be
    a. Placed in a special waste container
    b. Washed down the drain
    c. Placed in the wastebasket

Answer Key to Safety Quiz

Question	Answer	Question	Answer	Question	Answer
1	C	8	C	15	C
2	A	9	A	16	A
3	B	10	B	17	A, C
4	A, C	11	A	18	B
5	C	12	B	19	A, B, C
6	B	13	C	20	A
7	B	14	C		

Commitment to Safety in the Laboratory

I have read the laboratory preparation and safety procedures and agree that I will comply with the safety rules by carrying out the following:

_____ Read laboratory instructions ahead of lab time.

_____ Know the locations of eyewash fountains, fire extinguishers, safety showers, and exits.

_____ Wear safety goggles or safety glasses in the laboratory at all times.

_____ Use the proper equipment for laboratory procedures.

_____ Never perform any unauthorized experiments or work alone in the laboratory.

_____ Remember what is hot if I have used the Bunsen burner or hot plate.

_____ Clean up chemical spills and broken glass immediately.

_____ Immediately inform the instructor of a chemical spill or accident in the laboratory.

_____ Never eat or drink in the laboratory.

_____ Wear sensible clothing and closed shoes, and tie back long hair in the laboratory.

_____ Read the labels on reagent bottles carefully, remove only small amounts of reagent with the proper tools, and never return unused chemical to the bottle.

_____ Dispose of broken glass and waste chemicals in the appropriate waste container.

_____ Wash my hands and leave a clean work area when I leave the lab.

_____ Never leave an experiment when substances are heating or reacting.

_____

Your signature

_____   _____

Laboratory class and section   Date

F. **Laboratory Equipment**

When experiments call for certain pieces of equipment, it is important that you know which item to select. Using the wrong piece of equipment can lead to errors in measurement or cause a procedure to be done incorrectly. For safety and proper results, you need to identify the laboratory items you will use in doing lab work.

Look for each of the items shown in the pictures of laboratory equipment on pages xvi–xvii. The quiz will help you learn some of the common laboratory equipment.

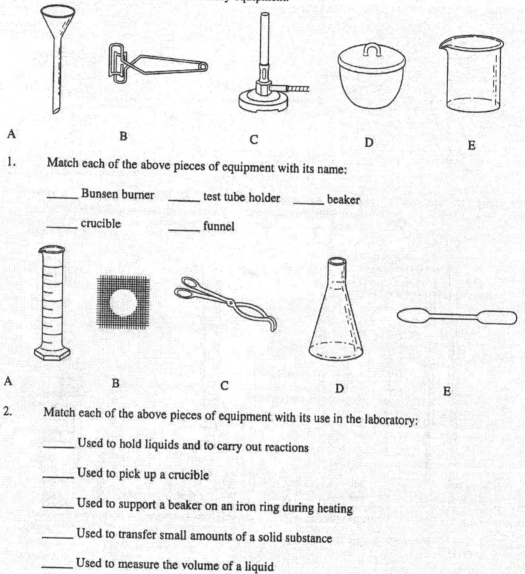

A   B   C   D   E

1. Match each of the above pieces of equipment with its name:

   ____ Bunsen burner   ____ test tube holder   ____ beaker

   ____ crucible   ____ funnel

A   B   C   D   E

2. Match each of the above pieces of equipment with its use in the laboratory:

   ____ Used to hold liquids and to carry out reactions

   ____ Used to pick up a crucible

   ____ Used to support a beaker on an iron ring during heating

   ____ Used to transfer small amounts of a solid substance

   ____ Used to measure the volume of a liquid

A Visual Guide to Laboratory Equipment

Evaporating dish
Used to evaporate a solution to dryness

Crucible and cover
Used to heat small samples to high temperatures

Crucible tongs
Used to pick up a crucible

Watch glass
Used to cover a beaker or to hold a small amount of a substance

Stirring rod
Used to mix or combine two or more substances in a test tube or a beaker

Forceps
Used to pick up a small object or one that is hot

Pinch clamp
Used to close rubber tubing

Spatula
Used to transfer small amounts of a solid

File
Used to cut glass tubing

Thermometer
Used to measure the temperature of a substance

Pipet
Used to transfer a specific volume of liquid solution to a container

Buret
Used to deliver a measured amount of solution with a known concentration

Medicine dropper
Used to deliver drops of a solution or liquid

A Visual Guide to Laboratory Equipment

Graphing Experimental Data

When a group of experimental quantities is determined, a graph can be prepared that gives a pictorial representation of the data. After a data table is prepared, a series of steps is followed to construct a graph.

Preparing a Data Table

A data table is prepared from measurements. Suppose we measured the distance traveled in a given time by a bicycle rider. Table 1 is a data table prepared by listing the two variables, time and distance, that we measured.

Table 1 *Time and Distance Measurement*

Time (hr)	Distance (km)
1	5
3	14
4	20
6	30
7	33
8	40
9	46
10	50

Constructing the Graph

Draw vertical and horizontal axes Draw a vertical and a horizontal axis on the appropriate graph paper. The lines should be set in to leave a margin for numbers and labels, but the graph should cover most of the graph paper. Place a title at the top of the graph. The title should describe the quantities that will be placed on the axes.

Label each axis The label for each axis reflects the measurement listed in the data table. On our sample graph, the labels are time (hr) for the horizontal axis and distance (km) for the vertical axis.

Apply constant scales On each axis apply a scale of equal intervals that includes the full range of data points (low to high) you have in the data table. The intervals on a scale must be equally spaced and fit on the line you have drawn. Do not exceed the graph lines. Use intervals on each axis that are convenient counting units (2, 4, 6, 8, etc. or 5, 10, 15, etc.). The interval size on one axis does not need to match the size of the intervals on the other axis.

For our sample graph, we used a scale for a distance range of 0 km to 50 km. Each graph division represents 5 km. (You only have to number a few lines in order to interpret the scale. It gets too crowded with numbers if every line is marked.) Every two divisions on the time scale represent a time interval of 1 hour within the 10-hour time span of the bicycle ride. See Figure 1.

Figure 1 Marking equal intervals for distance and time on the axes

Plot the data points Plot the points for each pair of measurements on the data table. Follow a measurement on the vertical axis across until it meets a line that would be drawn from the corresponding measured value on the horizontal axis.

For example, at 4 hours, the rider has traveled 20 km. On the graph, find 20 km on the distance scale, and 4 hr on the time scale. Then follow the perpendicular lines to where they intersect. That is a point on the graph. Plotting each data pair will show the relationship between distance and time. A smooth line or curve is drawn that best fits the data points. However, some points may not fit on the line or curve you draw. That occurs when error is associated with the measurements or when the data are affected by other variables, such as terrain and energy level for the bicycle rider. See Figure 2.

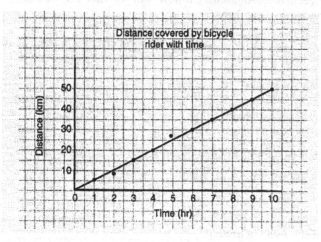

Figure 2 A completed graph with data points connected in a smooth line

Using the Laboratory Burner

EXPERIMENTAL PROCEDURES   **GOGGLES REQUIRED!**

Materials: Bunsen burner, striker or matches

In the laboratory, substances are often heated with a Bunsen burner, shown in the figure below. The burner consists of a metal tube and base connected to a gas source. The flow of gas is controlled by adjusting the gas lever at the bench or by turning the wheel at the base of the burner. The amount of air that enters the burner is adjusted by twisting the tube to open or close the air vents. The gas and air mixture is ignited at the top of the tube using a match or a striker. Make sure the gas valves are tightly closed when you leave the laboratory after using the Bunsen burner.

A typical laboratory burner

1. Before you light the burner, practice the following:
    a. Open and close the gas lever at the lab bench.
    b. Open and close the gas needle valve (wheel) at the base of the burner.
    c. Open and close the air vents.

With the air vents closed, ignite the burner with a striker or match. Your instructor may demonstrate the use of the striker. Turn on the gas and hold the flame or spark at the top rim of the burner tube.

2. If the flame is yellow and sooty, the gas mixture does not have an adequate supply of oxygen. Open the air vents until the color of the flame changes to blue. Adjust the gas flow until you have a flame that is 6–8 cm high with two distinct parts, an inner cone and an outer flame.

3. The hottest part of a flame is at the tip of the inner blue flame. For the most effective heating, be sure that the tip of the inner flame is placed just under the substance you heat. Remember what you heated: Hot metal and glass items do not look hot!

Questions

Q.1 What is the color of the flame with the air vent closed? _____

Open? _____

How do you control the height of the flame? _____

Q.2 What is the appearance of a flame that is used for heating? _____

Q.3 Where is the hottest part of a flame? _____

D1

Conversion Factors in Calculations

Goals

- Round off a calculated answer to the correct number of significant figures.
- Determine the area of a rectangle and the volume of a solid by direct measurement.
- Determine metric and metric-to-U.S.-unit equalities and corresponding conversion factors.
- Use conversion factors in calculations to convert units of length, volume, and mass.
- Convert a Celsius temperature to Fahrenheit and Kelvin.

Text Reference

Review the following before performing this laboratory.

Timberlake; *General, Organic, and Biological Chemistry: Structures of Life*, Third Edition
Sections 1.6; Pages 32 to 43

or

Timberlake; *Chemistry: An Introduction to General, Organic, and Biological Chemistry*, Tenth Edition
Sections 1.4 to 1.6; Pages 23 to 35 and Section 2.3; Pages 59 to 63

Discussion

As you begin to perform laboratory experiments, you will make measurements, collect data, and carry out calculations. When you use measured numbers in calculations, the answers that you report must reflect the precision of the original measurements. Thus it is often necessary to adjust the results you see on the calculator display. Every time you use your calculator, you will need to assess the mathematical operations, count significant figures, and round off calculator results.

A. Rounding Off

Usually there are fewer significant figures in the measured numbers used in a calculation than there are digits that appear in a calculator display. Therefore, we adjust the calculator result by rounding off. If the first number to be dropped is *less than 5,* it and all following numbers are dropped. If the first number to be dropped is *5 or greater,* the numbers are dropped and the value of the last *retained* digit is increased by 1. When you round a large number, the correct magnitude is retained by replacing the dropped digits with *placeholder zeros.* See Sample Problem D1.1. When a whole number appears in the calculator display, significant zeros may be added.

Laboratory D1

Sample Problem D1.1

Round off each of the following calculator displays to report an answer with three significant figures and another answer with two significant figures:

a. 75.6243   b. 0.528392   c. 387,600   d. 4

Solution:	**Three Significant Figures**	**Two Significant Figures**
a.	75.6	76
b.	0.528	0.53
c.	388,000	390,000
d.	4.00	4.0

B. Significant Figures in Calculations

When you carry out mathematical operations, the answer you report depends on the number of significant figures in the data you used.

Multiplication/division When you multiply or divide numbers, report the answer with the same number of significant figures as the measured number with the *fewest significant figures* (the least precise). See Sample Problem D1.2.

Sample Problem D1.2

Solve: $\dfrac{0.025 \times 4.62}{3.44} =$

Solution:

On the calculator, the steps are

$0.025 \times 4.62 \div 3.44$ = 0.033575581   *calculator display*
                            = 0.034   *final answer rounded to two significant figures*

Addition/subtraction When you add or subtract numbers, the reported answer has the same number of decimal places as the measured number with the *fewest decimal places*. See Sample Problem D1.3

Sample Problem D1.3

Add: 2.11 + 104.056 + 0.1205

Solution: 2.11   *two decimal places*
            104.056
            <u>0.1205</u>
            106.2865   *calculator display*
            106.29   *final answer rounded to two decimal places*

D-2

C. Conversion Factors for Length

Metric factors If a quantity is expressed in two different metric units, a *metric equality* can be stated. For example, the length of 1 meter is the same as 100 cm, which gives the *equality* 1 m = 100 cm. When the values in the equality are written as a fraction, the ratio is called a *conversion factor*. Two fractions are always possible, and both are conversion factors for the equality. These types of definitions are exact and do not limit the number of significant figures when used in a calculation.

Equality *Conversion Factors*

1 m = 100 cm $\dfrac{100 \text{ cm}}{1 \text{ m}}$ and $\dfrac{1 \text{ m}}{100 \text{ cm}}$

Metric–U.S. system factors When a quantity measured in a metric unit is compared to its measured quantity in a U.S. unit, a *metric–U.S.* conversion factor can be written. For example, 1 inch is the same length as 2.54 cm, as seen in Figure D1.1.

Figure D1.1 Comparing centimeters and inches

For the length of 1 inch, two conversion factors can be written.

Equality *Conversion Factors*

1 in. = 2.54 cm $\dfrac{2.54 \text{ cm}}{1 \text{ in.}}$ and $\dfrac{1 \text{ in.}}{2.54 \text{ cm}}$

D. Conversion Factors for Volume

In the metric system, equalities for volume can be written along with their corresponding conversion factors. A useful metric–U.S. equality is the relationship of 1 quart equaling 946 mL.

Laboratory D1

Equality	Conversion Factors
1 L = 1000 mL	$\dfrac{1000\ mL}{1\ L}$ and $\dfrac{1\ L}{1000\ mL}$
1 qt = 946 mL	$\dfrac{946\ mL}{1\ qt}$ and $\dfrac{1\ qt}{946\ mL}$

E. Conversion Factors for Mass

In the metric system, equalities for mass can be written along with their corresponding conversion factors. A useful metric–U.S. equality is the relationship of 454 g equaling 1 pound.

Equality	Conversion Factors
1 kg = 1000 g	$\dfrac{1000\ g}{1\ kg}$ and $\dfrac{1\ kg}{1000\ g}$
1 lb = 454 g	$\dfrac{454\ g}{1\ lb}$ and $\dfrac{1\ lb}{454\ g}$

F. Converting Temperature

Temperature measures the intensity of heat in a substance. A substance with little heat feels cold. Where the heat intensity is great, a substance feels hot. The temperature of our bodies is an indication of the heat produced. An infection may cause body temperature to deviate from normal. On the Celsius scale, water freezes at 0°C; on the Fahrenheit scale, water freezes at 32°F. A Celsius temperature is converted to its corresponding Fahrenheit temperature by using the following equation:

$$T_F = 1.8\,(T_C) + 32$$

When the Fahrenheit temperature is known, the Celsius temperature is determined by rearranging the equation. Be sure you subtract 32 from the T_F, then divide by 1.8.

$$T_C = \frac{(T_F - 32)}{1.8}$$

A Celsius temperature can be converted to a Kelvin temperature by using the following equation:

$$T_K = T_C + 273$$

Conversion Factors in Calculations

Lab Information

Time: 3 hr

Comments: Tear out the report sheets and place them beside the procedures.

Determine what the smallest lines of measurement are on each measuring tool you use.

Include an estimated digit for each measurement.

Round off the calculator answers to the correct number of significant figures.

Related Topics: Conversion factors, significant figures in mathematical operations, calculator use

Experimental Procedures

A. Rounding Off

Materials: Meterstick, solid

Section	Step	Procedure	Check when finished
A.1	1	**Rounding:** A student has rounded off some numbers. Determine whether the rounding was done correctly. If it is incorrect, state what the student needs to do and write the correctly rounded number.	

Section	Step	Procedure	Check when finished
A.2	1	**Area:** Determine the length (cm) and width (cm) of the sides of the rectangle on the report sheet.	
A.2	2	Obtain a second set of measurements from another student. Record.	
A.2	3	Calculate the area (cm^2) of the rectangle using your measurements and this formula: Area = L × W.	
A.2	4	Obtain the area calculated by the other student.	
A.2	5	Compare the calculated areas from both sets of measurements.	

Section	Step	Procedure	Check when finished
A.3	1	**Volume of a solid by direct measurement:** Obtain a solid object that has a regular shape, such as a cube, rectangular solid, or cylinder.	
A.3	2	Record its shape.	

Laboratory D1

Section	Step	Procedure	Check when finished
A.3	3	Use a meterstick to determine the dimensions of the solid in centimeters (cm).	
A.3	4	Use the appropriate formula from the following list to calculate the volume in cm³.	

Shape	Dimensions to Measure	Formulas for Volume
Cube	Length (L)	$V = L^3$
Rectangular solid	Length (L), width (W), height (H)	$V = L \times W \times H$
Cylinder	Diameter (D), height (H)	$V = \dfrac{\pi D^2 H}{4} = \dfrac{3.14 D^2 H}{4}$

B. Significant Figures in Calculations

Section	Step	Procedure	Check when finished
B.1	1	Solve the multiplication and division problems.	
B.1	2	Report your answers with the correct number of significant figures.	

Section	Step	Procedure	Check when finished
B.2	1	Solve the addition and subtraction problems.	
B.2	2	Report your answers with the correct number of significant figures.	

C. Conversion Factors for Length

Materials: Meterstick

Section	Step	Procedure	Check when finished
C.1	1	**Metric factors:** Observe the markings for millimeters on a meterstick.	
C.1	2	Write an equality that states the number of millimeters in 1 meter.	
C.1	3	Write two metric conversion factors for the relationship.	
C.1	4	Write equality and corresponding conversion factors for the relationship between centimeters and millimeters.	

Conversion Factors in Calculations

Section	Step	Procedure	Check when finished
C.2	1	**Metric–U.S. system factors:** Measure the length of the dark line on the report sheet in centimeters and in inches.	
C.2	2	Convert any fraction to a decimal number.	
C.2	3	Divide the number of centimeters by the number of inches to give a relationship.	
C.2	4	Round off correctly for your reported answer. This is your *experimental* value for the number of centimeters in 1 inch.	

Section	Step	Procedure	Check when finished
C.3	1	**Your metric height:** Record your height in inches. Or use a yardstick to measure.	
C.3	2	Using the appropriate conversion factors, *calculate* your height in centimeters and meters. Show your setup for each calculation.	

$$\text{Height (in.)} \times \frac{2.54 \text{ cm}}{1 \text{ in.}} = \text{your height (cm)}$$

$$\text{Height (cm)} \times \frac{1 \text{ m}}{100 \text{ cm}} = \text{your height (m)}$$

D. Conversion Factors for Volume

Materials: 1-L graduated cylinder, 1-quart measure (or two 1-pint measures)

Section	Step	Procedure	Check when finished
D.1	1	Observe the markings on a 1-liter graduated cylinder.	
D.1	2	Write an equality that states the number of milliliters in 1 liter.	
D.1	3	Write two conversion factors for the equality.	

Section	Step	Procedure	Check when finished
D.2	1	Using a 1-pint or 1-quart measure, transfer 1 quart of water to a 1-liter graduated cylinder.	
D.2	2	Record the number of milliliters in 1 quart.	
D.2	3	Write the equality that states the number of milliliters in a quart.	
D.2	4	Write two conversion factors for the equality.	

Laboratory D1

E. Conversion Factors for Mass

Materials: Commercial product with mass (weight) of contents given on label

Section	Step	Procedure	Check when finished
E.1	1	**Grams and pounds:** Labels on commercial products list the amount of the contents in both metric and U.S. units. Obtain a commercial product.	
E.1	2	Record the mass (weight) of the contents stated on the product label. Do not weigh.	
E.1	3	If the weight is given in ounces, convert it to pounds (1 lb = 16 oz)	
E.1	4	Divide the grams of the product by its weight in pounds. (Be sure to use the correct number of significant figures.) This is your value for grams in one pound (g/lb)	

_____ oz x $\dfrac{1 \text{ lb}}{16 \text{ oz}}$ = _____ lb

Section	Step	Procedure	Check when finished
E.2	1	**Pounds and kilograms:** State the mass of the label in kilograms. If necessary, convert the number of grams to kilograms.	
E.2	2	Divide the number of pounds by the number of kilograms. Report the ratio as lb/kg. (Be sure to use the correct number of significant figures.)	

_____ g x $\dfrac{1 \text{ kg}}{1000 \text{ g}}$ = _____ kg

Conversion Factors in Calculations

F. Converting Temperature

Materials: Thermometer (°C), a 150- or 250-mL beaker, ice, and rock salt

Section	Step	Procedure	Check when finished
F.1	1	Observe the markings on a thermometer.	
F.1	2	Indicate the lowest and highest temperatures that can be read using that thermometer. *Caution: Never shake down a laboratory thermometer. Shaking a laboratory thermometer can cause breakage and serious accidents.*	

Section	Step	Procedure	Check when finished
F.2	1	To measure the temperature of a liquid, place the bulb of the thermometer in the center of the solution. Keep it *immersed* while you read the temperature scale. When the temperature becomes constant, record the temperature (°C). On most thermometers, you can estimate the tenths of a degree (0.1°C). A set of beakers with the following contents may be set up in the lab; otherwise fill the beakers as instructed.	
F.2	2	Determine the temperature of room temperature by placing the thermometer on the lab bench.	
F.2	3	Record the temperature.	
F.2	4	Determine the temperature of tap water by filling a 250-mL beaker about 1/3 full of water.	
F.2	5	Record the temperature.	
F.2	6	Determine the temperature of an ice water mixture by adding enough ice to the tap water from step 4 to double (approximately) the volume. Allow 5 minutes for the temperature to change.	
F.2	7	Record the temperature.	
F.2	8	Determine the temperature of a salted ice mixture by adding rock salt to the ice water mixture in step 6. Stir and allow a few minutes for the temperature to change.	

Laboratory D1

Section	Step	Procedure	Check when finished
F.3	1	Convert the Celsius temperatures to corresponding temperature on the Fahrenheit and Kelvin scales.	

Report Sheet - Lab D1

Date _____ Name _____
Section _____ Team _____
Instructor _____   _____

Pre-Lab Study Questions

1. What are the rules for rounding off numbers?

2. How do you determine how many significant figures to keep in an answer obtained by multiplying or dividing?

3. How is the number of digits determined for an answer obtained by adding or subtracting?

4. Is a body temperature of 39.4°C a normal temperature or does it indicate a fever?

5. What is an equality and how is it used to write a conversion factor?

Laboratory D1

Conversion Factors in Calculations

Report Sheet - Lab D1

Date _____ Name _____
Section _____ Team _____
Instructor _____   _____

A. Rounding Off

A.1 **Rounding** A student rounded off the following calculator displays to three significant figures. Indicate if the rounded number is correct. If it is incorrect, round off the display value properly.

Calculator Display	Student's Rounded Value	Correct (yes/no)	Corrected (if needed)
24.4704	24.5		
143.63212	144		
532,800	530		
0.00858345	0.009		
8	8.00		

A.2 **Area**

	Your measurements	Another student's measurements
Length =	_____	_____
Width =	_____	_____
Area =	_____	_____

(*Show calculations.*)

Why could two students obtain different values for the calculated areas of the same rectangle?

Laboratory D1

Report Sheet - Lab D1

A.3 **Volume of a solid by direct measurement**

        Shape of solid   _____

Formula for volume of solid   _____

height _____   **length** _____

width _____   **diameter (if cylinder)** _____

Volume of the solid _____
(*Show calculations of volume including the units.*)

B. Significant Figures in Calculations

B.1 Perform the following multiplication and division calculations. Give a final answer with the correct number of significant figures:

       4.5×0.28 _____

       $0.1184 \times 8.00 \times 0.0345$ _____

       $\dfrac{(42.4)(15.6)}{1.265}$ _____

       $\dfrac{(35.56)(1.45)}{(4.8)(0.56)}$ _____

B.2 Perform the following addition and subtraction calculations. Give a final answer with the correct number of significant figures.

       13.45 mL + 0.4552 mL   _____

       145.5 m + 86.58 m + 1045 m   _____

       1315 + 200 + 1100   _____

       245.625 g − 80.2 g   _____

       4.62 cm − 0.885 cm   _____

Conversion Factors in Calculations

Report Sheet - Lab D1

Questions and Problems

Q.1   What is the combined mass in grams of objects that have masses of 0.2000 kg, 80.0 g, and 524 mg?

Q.2   A beaker has a mass of 225.08 g. When a liquid is added to the beaker, the combined mass is 238.254 g. What is the mass in grams of the liquid?

C. Conversion Factors for Length

C.1   **Metric factors**

Equality: 1 m   = _____ mm

Conversion factors: _____ m/mm   and   _____ mm/m

Equality: 1 cm = _____ mm

Conversion factors: _____ cm/mm   and   _____ mm/cm

C.2   **Metric–U.S. system factors**

Line length (*measured*)   _____ in.

                             _____ cm

_____ cm/in.   =   _____ cm / 1 in.   (*Experimental ratio*)

How close is your *experimental ratio* to the standard conversion factor of 2.54 cm/in.?

Laboratory D1

Report Sheet - Lab D1

C.3 **Your metric height**

Height (inches) _____

Height in centimeters *(calculated)*

_____ in. × $\dfrac{_____ \text{cm}}{1 \text{ in.}}$ = _____ cm

What is your height in meters? _____ m

(Show your calculations here.)

Questions and Problems *(Show complete setups.)*

Q.3 A pencil is 16.2 cm long. What is its length in millimeters (mm)?

Q.4 A roll of tape measures 45.5 inches. What is the length of the tape in meters?

D. Conversion Factors for Volume

D.1 Equality: 1 L = _____ mL

Conversion factors: $\dfrac{_____ \text{mL}}{\text{L}}$ and $\dfrac{_____ \text{L}}{\text{mL}}$

D.2 Number of milliliters in 1 quart: _____ mL/qt *(experimental)*

Equality: 1 qt = _____ mL

Conversion factors: $\dfrac{_____ \text{qt}}{\text{mL}}$ and $\dfrac{_____ \text{mL}}{\text{qt}}$

Conversion Factors in Calculations

Report Sheet - Lab D1

Questions and Problems *(Show complete setups.)*

Q.5 A patient received 825 mL of fluid in one day. What is that volume in liters?

Q.6 How many liters of plasma are present in 8.5 pints? (1 qt = 2 pt)

E. Conversion Factors for Mass

E.1 **Grams and pounds**

Name of commercial product _____

Mass in grams stated on label _____

Weight in lb or oz given on label _____

Weight in lb _____
(Convert oz to lb if needed.)

<u>Number of grams</u>=   _____ g   =   _____ g
  Number of lb   lb   1 lb

How does your experimental factor compare to the standard value of 454 g/lb?

E.2 **Pounds and kilograms**

Mass in kilograms (from label) _____

Weight in lb _____

<u>Number of lbs</u>=   _____ lb   =   _____ lb
Number of kg   kg   1 kg

Laboratory D1

Report Sheet - Lab D1

How does your *experimental factor* compare to the standard value of 2.20 lb/kg?

Questions and Problems

Q.7 An infant has a mass of 3.40 kg. What is the weight of the infant in pounds?

F. Converting Temperature

F.1 Temperature scale(s) on the laboratory thermometer _____

Lowest temperature _____ Highest temperature _____

	(G.2) °C	(G.3) °F	K
a. Room temperature			
b. Tap water			
c. Ice–water mixture			
d. Salt ice–water mixture			

Questions and Problems

Q.8 Write an equation for each of the following temperature conversions:

a. °C to °F

b. °F to °C

c. °C to K

Q.9 A recipe calls for a baking temperature of 205°C. What temperature in °F should be set on the oven?

D2

Compounds and Their Formulas

Goals

- Compare physical properties of a compound with the properties of the elements that formed it.
- Identify a compound as ionic or covalent.
- Determine the subscripts in the formula of a compound.
- Write the electron-dot structure for an atom and an ion.
- Write a correct formula and name of an ionic or covalent compound.
- Write a correct formula and name of a compound containing a polyatomic ion.

Text Reference

Review the following before performing this laboratory.

Timberlake; *General, Organic, and Biological Chemistry: Structures of Life*, Third Edition
Sections 5.1 to 5.9; Pages 168 to 202

or

Timberlake; *Chemistry: An Introduction to General, Organic, and Biological Chemistry*, Tenth Edition
Sections 4.1 to 4.8; Pages 121 to 153

Discussion

Nearly everything is made of compounds. A compound consists of two or more different elements that are chemically combined. Most atoms form compounds by forming octets in their outer shells. The attractions between the atoms are called chemical bonds. For example, when a metal combines with a nonmetal, the metal loses electrons to form a positive ion and the nonmetal gains electrons to form a negative ion. The attraction between the positive ions and the negative ions is called an ionic bond. When two nonmetals form a compound, they share electrons and form covalent bonds. In covalent compounds, the atoms are bonded as individual units called molecules. See Table D2.1.

Table D2.1 Types of Bonding in Compounds

Compound	Types of Elements	Characteristics	Type of Bonding
NaCl	Metal, nonmetal	Ions (Na$^+$, Cl$^-$)	Ionic
MgBr$_2$	Metal, nonmetal	Ions (Mg^{2+}, Br$^-$)	Ionic
CCl$_4$	Two nonmetals	Molecules	Covalent
NH$_3$	Two nonmetals	Molecules	Covalent

Laboratory D2

In a compound, there is a definite proportion of each element. This is represented in the formula, which gives the lowest whole number ratio of each kind of atom. For example, water has the formula H_2O. This means that two atoms of hydrogen and one atom of oxygen are combined in every molecule of water. Water never has any other formula.

When we observe a compound or an element, we see physical properties such as color and luster. We measure other physical properties such as density, melting point, and boiling point. When elements undergo chemical combination, the physical properties change to the physical properties of the new substances that form. For example, when silver tarnishes, the physical property of the shiny, silver metal changes to the dull, gray color as silver combines with sulfur to form tarnish, Ag_2S. A chemical change has occurred when the reaction between elements causes a change in their physical properties.

A. Electron-Dot Structures

When atoms of metals in Groups 1A, 2A, or 3A react with atoms of nonmetals in Groups 5A, 6A, or 7A, the metals lose electrons and the nonmetals gain electrons in their valence shells. We can predict the number of electrons lost or gained by analyzing the electron-dot structures of the atoms. In an electron-dot structure, the valence electrons are represented as dots around the symbol of the atom. For example, calcium, electron arrangement 2-8-8-2, has two valence electrons and an electron-dot structure with two dots. Chlorine, electron arrangement 2-8-7, has seven valence electrons and an electron-dot structure with seven dots.

$\cdot \overset{\cdot}{Ca} \cdot \qquad \cdot \overset{\cdot \cdot}{\underset{\cdot \cdot}{Cl}} \colon$

Ca loses two electrons to attain an octet. This gives it an ionic charge of 2+. It is now a calcium ion with an electron arrangement of 2-8-8. As a positive ion, it keeps the same name as the element.

	Calcium Atom, Ca	**Calcium Ion, Ca^{2+}**	
Electron arrangement	2-8-8-2	2-8-8	(Two electrons lost)
Number of protons	$20p^+$	$20p^+$	(Same)
Number of electrons	$20e^-$	$18e^-$	(Two fewer electrons)
Net ionic charge	0	2+	

When nonmetals (5, 6, or 7 valence electrons) combine with metals, they gain electrons to become stable, and form negatively charged ions. For example, a chlorine atom gains one valence electron to become stable with an electron arrangement of 2-8-8. With the addition of one electron, chlorine becomes a chloride ion with an ionic charge of 1–. In the name of a binary compound with two different elements, the name of the negative ion ends in "-ide".

	Chlorine Atom, Cl	Chloride Ion, Cl⁻	
Electron arrangement	2-8-7	2-8-8	(Electron added)
Number of protons	$17p^+$	$17p^+$	(Same)
Number of electrons	$17e^-$	$18e^-$	(One more electron)
Net ionic charge	0	1–	

B. Ionic Compounds and Formulas

The group number on the periodic table can be used to determine the ionic charges of elements in each family of elements. Nonmetals form ions when they combine with a metal.

Group number	1A	2A	3A	4A	5A	6A	7A	8
Valence electrons	$1e^-$	$2e^-$	$3e^-$	$4e^-$	$5e^-$	$6e^-$	$7e^-$	$8e^-$
Electron change	lose 1	lose 2	lose 3	none	gain 3	gain 2	gain 1	no change
Ionic charge	1+	2+	3+	none	3–	2–	1–	none

In an ionic formula, the number of electrons lost is equal to the number of electrons gained. The overall net charge is zero. To balance the charge, we must determine the smallest number of positive and negative ions that give an overall charge of zero (0). We can illustrate the process by representing the ions Ca2+ and Cl– as geometric shapes.

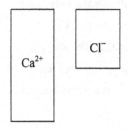

The charge is balanced by using two Cl⁻ ions to match the charge of the Ca^{2+} ion. The number of ions needed gives the subscripts in the formula for the compound $CaCl_2$. (The subscript 1 for Ca is understood.) In any ionic formula, only the symbols are written, not their ionic charges.

Ca^{2+} | Cl⁻ / Cl⁻ | = $CaCl_2$

Laboratory D2

C. Ionic Compounds with Transition Metals

Most of the transition metals can form more than one kind of positive ion. We will illustrate variable valence with iron. Iron forms two ions, one (Fe^{2+}) with a 2+ charge, and another (Fe^{3+}) with a 3+ charge. To distinguish between the two ions, a Roman numeral that gives the ionic charge of that particular ion follows the element name. The Roman numeral is always included in the names of compounds with variable positive ions. In an older naming system, the ending "-ous" indicates the lower valence; the ending "-ic" indicates the higher one. See Table D2.2.

Table D2.2 *Some Ions of the Transition Elements*

Ion	Names	Compound	Names
Fe^{2+}	Iron(II) ion or ferrous ion	$FeCl_2$	Iron(II) chloride or ferrous chloride
Fe^{3+}	Iron(III) ion or ferric ion	$FeCl_3$	Iron(III) chloride or ferric chloride
Cu^+	Copper(I) ion or cuprous ion	$CuCl$	Copper(I) chloride or cuprous chloride
Cu^{2+}	Copper(II) ion or cupric ion	$CuCl_2$	Copper(II) chloride or cupric chloride

Among the transition metals, a few elements (zinc, silver, and cadmium) form only a single type of ion; they have a fixed ionic charge. Thus, they are not variable and do not need a Roman numeral in their names.

Zn^{2+} zinc ion $\qquad$ Ag^+ silver ion $\qquad$ Cd^{2+} cadmium ion

D. Ionic Compounds with Polyatomic Ions

A compound that consists of three or more kinds of atoms will contain a polyatomic ion. A polyatomic ion is a group of atoms with an overall charge. That charge, which is usually negative, is the result of adding electrons to a group of atoms to complete octets. The most common polyatomic ions consist of the nonmetals C, N, S, P, Cl, or Br combined with two to four oxygen atoms. Some examples are given in Table D2.3. The ions are named by replacing the ending of the nonmetal with "-ate" or "-ite". The "-ite" ending has one oxygen less than the most common form of the ion, which has an "-ate" ending. Ammonium ion, NH_4^+, is positive because its group of atoms lost one electron.

Table D2.3 *Some Polyatomic Ions*

Common Polyatomic Ion		One Oxygen Less	
NH_4^+	ammonium ion		
OH^-	hydroxide ion		
NO_3^-	nitrate ion	NO_2^-	nitrite ion
CO_3^{2-}	carbonate ion		
HCO_3^-	bicarbonate ion (hydrogen carbonate ion)		
SO_4^{2-}	sulfate ion	SO_3^{2-}	sulfite ion

Compounds and Their Formulas

HSO_4^-	bisulfate ion (hydrogen sulfate ion)	HSO_3^-	bisulfite ion (hydrogen sulfite ion)
PO_4^{3-}	phosphate ion	PO_3^{3-}	phosphite ion

To write a formula with a polyatomic ion, we determine the ions needed for charge balance just as we did with the simple ions. When two or more polyatomic ions are needed, the formula of the ion is enclosed in parentheses and the subscript placed outside. No change is ever made in the formula of the polyatomic ion itself. Consider the formula of the compound formed by Ca^{2+} and NO^{3-} ions.

E. Covalent (Molecular) Compounds

Covalent bonds form between two nonmetals found in Groups 4A, 5A, 6A, or 7A or 8A. In a covalent compound, octets are achieved by sharing electrons between atoms. The sharing of one pair of electrons is called a single bond. A double bond is the sharing of two pairs of electrons between atoms. In a triple bond, three pairs of electrons are shared. To write the formula of a covalent compound, determine the number of electrons needed to complete an octet. For example, nitrogen in Group 5A has five valence electrons. Nitrogen atoms need three more electrons for an octet; they share three electrons.

Electron-Dot Structures

The formulas of covalent compounds are determined by sharing the valence electrons until each atom has an octet. For example, in water (H_2O), oxygen shares two electrons with two hydrogen atoms. Oxygen has an octet and hydrogen is stable because it has two electrons in the first valence shell.

Dot Structure for H_2O

In another example, we look at a compound, CO_2, that has double bonds. In the elements' electron-dot structures, carbon has 4 valence electrons and each oxygen atom has 6. Thus a total of 16 (4 + 6 + 6) electrons can be used in forming the octets by sharing electrons. We can use the following steps to determine the electron-dot structure for CO_2:

Laboratory D2

1. Connect the atoms with pairs of electrons, thus using 4 electrons.

O : C : O

2. Place the remaining 12 electrons (16 – 4) around the atoms. Don't add more electrons.

:Ö: C :Ö:

3. If octets *cannot* be completed, try sharing more electrons. In step 2, the octets are complete for the oxygen atoms, but not for the carbon. One pair of electrons from each oxygen atom is moved to share with carbon. Now all the atoms have octets. There are still 16 electrons used, but they are now arranged to give each atom an octet. There are two double bonds in the CO_2 molecule.

sharing two pairs of electrons makes double bonds

Names of Covalent Compounds

Binary (two-element) covalent compounds are named by using prefixes that give the number of atoms of each element in the compound. The first nonmetal is named by the element name; the second ends in ide. The prefixes are derived from the Greek names: mono (1), di (2), tri (3), tetra (4), penta (5), hexa (6), hepta (7), and octa (8). Usually the prefix mono is not shown for the first element. See Table D2.4.

Table D2.4 *Some Formulas and Names of Covalent Compounds*

Formula	Name
CO	carbon **mono**xide
CO_2	carbon **di**oxide
PCl_3	phosphorus **tri**chloride
N_2O_4	**di**nitrogen **tetr**oxide (drop *a* in a double vowel)
SCl_6	sulfur **hexa**chloride

Compounds and Their Formulas

Lab Information

Time:	2–3 hr
Comments:	Tear out the report sheets and place them beside the matching procedures.
Related Topics:	Ions, ionic bonds, naming ionic compounds, covalent bond, covalent compounds, naming covalent compounds

Experimental Procedures

A.   Electron-Dot Structures

Section	Step	Procedure	Check when finished
A.1	1	Write the electron arrangements for atoms and their ions.	
A.1	2	Determine the number of electrons lost or gained.	
A.1	3	Write the electron-dot structure of the ion that each would form along with its symbol, ionic charge, and name.	

B.   Ionic Compounds and Formulas

Materials: Reference books: *Merck Index* or *CRC Handbook of Chemistry and Physics,* display of compounds

Section	Step	Procedure	Check when finished
B.1	1	**Physical properties:** In the laboratory display of compounds observe NaCl, sodium chloride.	
B.1	2	Describe its appearance.	
B.1	3	Using a chemistry reference such as the *Merck Index* or the *CRC Handbook of Chemistry and Physics,* record the density and melting point.	

Section	Step	Procedure	Check when finished
B.2	1	**Formulas of ionic compounds:** Use the periodic table to write the positive and negative ion in each compound.	
B.2	2	Use charge balance (net total = zero) to write the correct formula.	
B.2	3	Use subscripts when two or more ions are needed	

D-25

Laboratory D2

Section	Step	Procedure	Check when finished
B.3	1	**Names of ionic compounds:** From the formula of each ionic compound, write the compound name by placing the metal name first, then the nonmetal name ending in "-ide".	

C. Ionic Compounds with Transition Metals

Section	Step	Procedure	Check when finished
C.1	1	**Physical properties:** In the display of compounds observe $FeCl_3$, iron(III) chloride or ferric chloride. Describe its appearance.	
C.1	2	Using a chemistry reference such as the *Merck Index* or the *CRC Handbook of Chemistry and Physics*, record the density and melting point.	

Section	Step	Procedure	Check when finished
C.2	1	**Formulas of ionic compounds:** Use the periodic table to write the positive and negative ion in each compound.	
C.2	2	Use charge balance (net total = zero) to write the correct formula. Use subscripts when two or more ions are needed.	

Section	Step	Procedure	Check when finished
C.3	1	**Names of ionic compounds:** From the formula of each ionic compound, write the compound name by placing the metal name first, then the nonmetal name ending in "-*ide*". Be sure to indicate the ionic charge if the transition metal has a variable valence by using a Roman numeral or using the "-*ous*" or "-*ic*" ending.	

D. Ionic Compounds with Polyatomic Ions

Section	Step	Procedure	Check when finished
D.1	1	**Physical properties:** In the display of compounds observe K_2CO_3, potassium carbonate. Describe its appearance.	
D.1	2	Using a chemistry reference book such as the *Merck Index* or the *CRC Handbook of Chemistry and Physics*, record the density and melting point.	

Section	Step	Procedure	Check when finished
D.2	1	**Formulas of ionic compounds:** Use the periodic table to write the positive and negative (polyatomic) ion in each compound.	
D.2	2	Use charge balance (net total = zero) to write the correct formula. Use subscripts when two or more ions are needed. Use parentheses when two or more polyatomic ions are needed for charge balance.	

Section	Step	Procedure	Check when finished
D.3	1	**Names of ionic compounds:** Name the compounds listed, using the correct names of the polyatomic ions.	

E. Covalent (Molecular) Compounds

Section	Step	Procedure	Check when finished
E.1	1	**Electron-dot formulas of elements:** Write the electron-dot structure for each nonmetal.	

Section	Step	Procedure	Check when finished
E.2	1	**Physical properties:** In the display of compounds observe water, H_2O. Describe its appearance.	
E.2	2	Using a chemistry reference such as the *Merck Index* or the *CRC Handbook of Chemistry and Physics*, record the density and melting point.	

Laboratory D2

Section	Step	Procedure	Check when finished
E.3	1	**Electron-dot structures:** Write the electron-dot structure for each covalent compound.	
E.3	2	Name each compound, using prefixes to indicate the number of atoms of each element. By convention, the prefix mono can be omitted from the name of the first nonmetal.	

Compounds and Their Formulas

Report Sheet - Lab D2

Date _____ Name _____
Section _____ Team _____
Instructor _____

Pre-Lab Study Questions

1. Where are the valence electrons in an atom?

2. How are positive and negative ions formed?

3. How do subscripts represent the charge balance of ions?

4. Why are electrons shared in covalent compounds?

Laboratory D2

5. How do the names of covalent compounds differ from the names of ionic compounds?

6. What are polyatomic ions?

Compounds and Their Formulas

Report Sheet - Lab D2

Date _____ Name _____
Section _____ Team _____
Instructor _____   _____

A. Electron-Dot Structures

Element	Atomic Number	Electron Arrangement of Atom	Electron-Dot Structure	Loss or Gain of Electrons	Electron Arrangement of Ion	Ionic Charge	Symbol of Ion	Name of Ion
Sodium	11	2-8-1	Na·	lose 1e^-	2-8	1+	Na^+	sodium ion
Nitrogen	7	2-5	·N̈·	gain 3e^-	2-8	3−	N^{3-}	nitride ion
Aluminum								
Chlorine								
Calcium								
Oxygen								

B. Ionic Compounds and Formulas

B.1 Physical properties

Compound	Appearance	Density	Melting Point

B.2 Formulas of ionic compounds

Name	Positive Ion	Negative Ion	Formula
Sodium chloride	Na^+	Cl^-	
Magnesium chloride			
Calcium oxide			
Lithium phosphide			
Aluminum sulfide			
Calcium nitride			

Laboratory D2

Report Sheet - Lab D2

B.3 Names of ionic compounds

K₂S	Potassium sulfide
BaF₂	
MgO	
Na₃N	
AlCl₃	
Mg₃P₂	

C. Ionic Compounds with Transition Metals

C.1 Physical properties

Compound	Appearance	Density	Melting Point

C.2 Formulas of ionic compounds

Name	Positive Ion	Negative Ion	Formula
Iron(III) chloride	Fe³⁺	Cl⁻	
Iron(II) oxide			
Copper(I) sulfide			
Copper(II) nitride			
Zinc oxide			
Silver sulfide			

C.3 Names of ionic compounds

Cu₂S	Copper(I) sulfide
Fe₂O₃	
CuCl₂	
FeS	
Ag₂O	
FeBr₂	

Compounds and Their Formulas

Report Sheet - Lab D2

D. Ionic Compounds with Polyatomic Ions

D.1 **Physical properties**

Compound	Appearance	Density	Melting Point

D.2 **Formulas of ionic compounds**

Name	Positive Ion	Negative Ion	Formula
Potassium carbonate	K^+	CO_3^{2-}	
Sodium nitrate			
Calcium bicarbonate			
Aluminum hydroxide			
Lithium phosphate			
Potassium sulfate			

D.3 **Names of ionic compounds**

$CaSO_4$	Calcium sulfate
$Al(NO_3)_3$	
Na_2CO_3	
$MgSO_3$	
$Cu(OH)_2$	
$Mg_3(PO_4)_2$	

Questions and Problems

Q.1 Write the correct formulas for the following ions:

   sodium ion _____   oxide ion _____   calcium ion _____

   chloride ion _____   sulfate ion _____   iron(II) ion _____

D-33

Laboratory D2

Report Sheet - Lab D2

E. Covalent (Molecular) Compounds

E.1 Electron-dot formulas of elements

Hydrogen	Carbon	Nitrogen	Oxygen	Sulfur	Chlorine
H·					

E.2 Physical properties

Compound	Appearance	Density	Melting Point

E.3 Electron-dot structures

Compound	Electron Dot Structure	Name
H_2O		
SBr_2		
PCl_3		
CBr_4		
SO_3		

Questions and Problems

Q.2   a. Identify each of the following compounds as ionic or covalent.

        b. Write the correct formula for each.

	Ionic/Covalent	Formula
sodium oxide		
iron(III) bromide		
sodium carbonate		
carbon tetrachloride		
nitrogen tribromide		

D3

Properties and Structures of Alkanes

Goals

- Observe chemical and physical properties of organic and inorganic compounds.
- Draw formulas for alkanes from their three-dimensional models.
- Write the names of alkanes from their structural formulas.
- Construct models of isomers of alkanes.
- Write the structural formulas for cycloalkanes.

Text References

Review the following before performing this laboratory.

Timberlake; *General, Organic, and Biological Chemistry: Structures of Life*, Third Edition
Sections 11.1 to 11.5; Pages 416 to 438

or

Timberlake; *Chemistry: An Introduction to General, Organic, and Biological Chemistry*, Tenth Edition
Sections 10.1 to 10.5; Pages 351 to 372

Discussion

A. Color, Odor, and Physical State

Organic compounds are made of carbon and hydrogen, and sometimes oxygen and nitrogen. Of all the elements, only carbon atoms bond to many more carbon atoms, a unique ability that gives rise to many more organic compounds than all the inorganic compounds known today. The covalent bonds in organic compounds and the ionic bonds in inorganic compounds account for several of the general differences we will observe in their physical and chemical properties. See Table D3.1.

Table D3.1 *Comparing Some Properties of Organic and Inorganic Compounds*

Organic Compounds	Inorganic Compounds
Covalent bonds	Ionic or polar bonds
Soluble in nonpolar solvents, not water	Soluble in water
Low melting and boiling points	High melting and boiling points
Strong, distinct odors	Usually no odor
Flammable	Not flammable

Laboratory D3

B. Solubility

Typically, inorganic compounds that are ionic are soluble in water, a polar compound, but organic compounds are nonpolar and thus are not soluble in water. However, organic compounds are soluble in organic solvents because they are both nonpolar. A general rule for solubility is that "like dissolves like."

C. Combustion

Many organic compounds react with oxygen, a reaction called combustion, to form carbon dioxide and water. Combustion is the reaction that occurs when gasoline burns with oxygen in the engine of a car or when natural gas, methane, burns in a heater or stove. In a combustion reaction, heat is given off; the reaction is exothermic. Equations for the combustion of methane and propane are written as follows:

$$CH_4(g) + 2O_2(g) \rightarrow CO_2(g) + 2H_2O(g) + heat$$
methane

$$C_3H_8(g) + 5O_2(g) \rightarrow 3CO_2(g) + 4H_2O(g) + heat$$
propane

D. Structures of Alkanes

The saturated hydrocarbons represent a group of organic compounds composed of carbon and hydrogen. Alkanes and cycloalkanes are called *saturated* hydrocarbons because their carbon atoms are connected by only single bonds. In each type of alkane, each carbon atom has four valence electrons and must always have four single bonds.

To learn more about the three-dimensional structure of organic compounds, it is helpful to build models using a ball-and-stick model kit or other types of model kits. In the kit are wooden (or plastic) balls, which represent the typical elements in organic compounds. Each wooden atom has the correct number of holes drilled for bonds that attach to other atoms. See Table D3.2. (Note, different model kits may have different color balls.)

Table D3.2 *Elements and Bonds Represented in the Organic Model Kit*

Color	Element	Number of Bonds
Black	carbon	4
Yellow	hydrogen	1
Red	oxygen	2
Green	chlorine	1
Orange	bromine	1
Purple	iodine	1
Blue	nitrogen	3
Bonds		
Sticks, springs		

Properties and Structures of Alkanes

The first model to build is methane, CH_4, a hydrocarbon consisting of one carbon atom and four hydrogen atoms. The model of methane shows the three-dimensional shape, a tetrahedron, around a carbon atom.

Three-dimensional structure   Complete structural formula   Condensed structural formula

To represent this model on paper, its shape is flattened, and the carbon atom is shown attached to four hydrogen atoms. This type of formula is called a *complete structural formula*. However, it is more convenient to use a shortened version called a *condensed structural formula*. To write a condensed formula, the hydrogen atoms are grouped with their carbon atom. The number of hydrogen atoms is written as a subscript. The complete structural formula and the condensed structural formula for C_2H_6 are shown below:

Complete structural formula   Condensed structural formula

Names of Alkanes

The names of alkanes all end with "*-ane*". The names of organic compounds are based on the names of the alkane family. See Table D3.3.

Table D3.3 *Names and Formulas of the First Ten Alkanes*

Name	Formula	Name	Formula
Methane	CH_4	Hexane	$CH_3CH_2CH_2CH_2CH_2CH_3$
Ethane	CH_3CH_3	Heptane	$CH_3CH_2CH_2CH_2CH_2CH_2CH_3$
Propane	$CH_3CH_2CH_3$	Octane	$CH_3CH_2CH_2CH_2CH_2CH_2CH_2CH_3$
Butane	$CH_3CH_2CH_2CH_3$	Nonane	$CH_3CH_2CH_2CH_2CH_2CH_2CH_2CH_2CH_3$
Pentane	$CH_3CH_2CH_2CH_2CH_3$	Decane	$CH_3CH_2CH_2CH_2CH_2CH_2CH_2CH_2CH_2CH_3$

E. Constitutional Isomers

Constitutional isomers are present when a molecular formula can represent two or more different structural (or condensed) formulas. One structure cannot be converted to the other without breaking and forming new bonds. The isomers have different physical and chemical properties. One of the reasons for the vast array of organic compounds is the phenomenon of isomerism.

Laboratory D3

Isomers of C₄H₁₀

```
    H  H  H  H
    |  |  |  |
H — C — C — C — C — H
    |  |  |  |
    H  H  H  H
       Butane
```

```
              H
              |
          H — C — H   ← A branch to a three-carbon chain
              |
       H      H
       |      |
   H — C — C — C — H
       |      |
       H      H  H
        2-Methylpropane
```

F. Cycloalkanes

In a cycloalkane, an alkane has a cyclic or ring structure. The structural formula of a cycloalkane indicates all of the carbon and hydrogen atoms. The condensed formula groups the hydrogen atoms with each of the carbon atoms. Another type of notation called the *geometric* structure is often used to depict a cycloalkane by showing only the bonds that outline the geometric shape of the compound. For example, the geometric shape of cyclopropane is a triangle, and the geometric shape of cyclobutane is a square. Examples of the various structural formulas for cyclobutane are shown below.

Complete structural formula   Condensed formula   Line-bond formula

D-38

Properties and Structures of Alkanes

Lab Information

Time: 2 hr

Comments: Tear out the report sheets and place them next to the matching procedures.

Related Topics: Organic compounds, hydrocarbons, solubility, combustion, complete structural formula, functional groups

A. Color, Odor, and Physical State *(This is a lab display.)*

Materials: Test tubes (6), test tube rack, spatulas, NaCl(s), KI(s), toluene, benzoic acid, cyclohexane water, chemistry handbook

Section	Step	Procedure	Check when finished
A	1	Observe the samples in a test tube rack in the hood (NaCl, KI, benzoic acid, cyclohexane, toluene, and water).	
A	2	Record the formula for each.	
A	3	Record the physical state (solid, liquid or gas).	
A	4	Record the odor of each by first taking a breath and hold it while you gently fan the air above the test tube toward you. Your instructor will demonstrate this technique.	
A	5	Look up the melting point of each compound using a chemistry handbook.	
A	6	Record.	
A	7	State the types of bonds in each as ionic or covalent.	
A	8	Identify each as an organic or inorganic compound.	

B. Solubility *(This is a lab display.)*

Materials: Test tubes, spatulas, NaCl(s), toluene, cyclohexane

Section	Step	Procedure	Check when finished
B	1	Observe a sample in the hood which has 10 drops of cyclohexane and 10 drops of water in a test tube.	
B	2	Record your observations.	
B	3	Identify the upper layer and the lower layer.	
B	4	Observe a test tube that has a few crystals of NaCl and a few drops of cyclohexane, a nonpolar solvent.	

Laboratory D3

B	5	Record whether NaCl is soluble (S) or insoluble (I) in cyclohexane.	
B	6	Observe a test tube which has 10 drops of toluene and 15 drops of cyclohexane, a nonpolar solvent.	
B	7	Record whether toluene is soluble (S) or insoluble (I) in cyclohexane.	
B	8	Observe a test tube that has a few crystals of NaCl and a few drops of water, a polar solvent.	
B	9	Record whether NaCl is soluble (S) or insoluble (I) in water.	
B	10	Observe a test tube which has 10 drops of toluene and 15 drops of water, a polar solvent.	
B	11	Record whether toluene is soluble (S) or insoluble (I) in water.	

C. Combustion *(This is a demonstration by your instructor.)*

Materials: 2 evaporating dishes, spatulas, wood splints, NaCl(*s*), cyclohexane

Section	Step	Procedure	Check when finished
C	1	Observe your instructor ignite a splint and hold the flame to a sample of NaCl.	
C	2	Record whether the NaCl burns.	
C	3	If the substance burns, note the color of the flame.	
C	4	Identify each as an organic or inorganic compound.	
C	5	Observe your instructor ignite a splint and hold the flame to a sample of cyclohexane.	
C	6	Record whether the cyclohexane burns.	
C	7	If the substance burns, note the color of the flame.	
C	8	Identify cyclohexane as an organic or inorganic compound.	

Properties and Structures of Alkanes

D. Structures of Alkanes

Materials: Organic model kit

Section	Step	Procedure	Check when finished
D.1	1	Using an organic model kit, construct a ball-and-stick model of a molecule of methane, CH_4. Place wooden dowels in all the holes in the carbon atom (black). Attach hydrogen (yellow) atoms to each. (Note model kits may have different colors.)	
D.1	2	Draw the three-dimensional (tetrahedral) shape of methane.	
D.1	3	Write the complete structural formula and the condensed structural formula of methane.	
D.2	1	Make a model of ethane, C_2H_6.	
D.2	2	Observe that the tetrahedral shape is maintained for each carbon atom in the structure.	
D.2	3	Write the complete structural and condensed structural formulas for ethane.	
D.3	1	Make a model of propane, C_3H_8.	
D.3	2	Write the complete structural and condensed structural formulas for propane.	

E. Constitutional Isomers

Materials: Organic model kit, chemistry handbook

Section	Step	Procedure	Check when finished
E.1	1	The molecular formula of butane is C_4H_{10}. Construct a model of butane by connecting four carbon atoms in a chain.	
E.1	2	Draw its complete and condensed structural formulas.	
E.2	1	Make an isomer of C_4H_{10} by switching an end $-CH_3$ group with a hydrogen atom on the center carbon.	
E.2	2	Complete the end of the chain with a hydrogen atom.	
E.2	3	Write its complete and condensed structural formulas.	
E.3	1	Obtain a chemistry handbook.	
E.3	2	For each isomer, find the molar mass, melting point, boiling point, and density.	
E.4	1	Make models of the three isomers of C_5H_{12}. Make the continuous-chain isomer first.	

Laboratory D3

E.4	2	Draw the complete structural and condensed structural formulas for each.	
E.4	3	Name each isomer.	
E.5	1	In the chemistry handbook, find the molar mass, melting point, boiling point, and density of each structural isomer.	

F. Cycloalkanes

Materials: Organic model kit

Section	Step	Procedure	Check when finished
F.1	1	Use the springs or flexible plastic in the model kits to make a model of the cycloalkane with three carbon atoms.	
F.1	2	Write the complete structural and condensed structural formulas.	
F.1	3	Draw the line-bond formula and name this compound.	
F.2	1	Use the springs in the model kits to make models of a cycloalkane with four carbon atoms.	
F.3	1	Use the springs or flexible plastic in the model kits to make models of a cycloalkane with five carbon atoms.	
F.3	2	Draw the complete structural and condensed structural formulas.	
F.3	3	Draw the line-bond formula and give the name for each.	

Properties and Structures of Alkanes

Report Sheet - Lab D4

Date _____ Name _____
Section _____ Team _____
Instructor _____   _____

Pre-Lab Study Questions

1. Would you expect an organic compound to be soluble in water? Why?

2. Which is more flammable: an organic or inorganic compound?

3. Do all organic compounds contain carbon?

4. A diamond is a crystalline form of carbon. Do you think that it is an organic compound?

5. What elements are present in alkanes?

Laboratory D3

6. How does a complete structural formula differ from a condensed structural formula?

7. If isomers of an alkane have the same molecular formula, how do they differ?

8. Why does a 4 carbon linear alkane, butane C_4H_{10}, have two fewer hydrogens atoms than a 4 carbon cyclic alkane, cyclobutane, C_4H_8.

Report Sheet - Lab D3

Date _____ Name _____
Section _____ Team _____
Instructor _____

A. Color, Odor, and Physical State

Name	Formula	Physical State	Odor	Melting Point	Type of Bonds?	Organic or Inorganic?
Sodium chloride						
Cyclohexane	C_6H_{12}					
Potassium iodide						
Benzoic acid	$C_7H_6O_2$					
Toluene	C_7H_8					
Water						

B. Solubility

In the mixture, water is the _____ layer and cyclohexane is the _____ layer.

Solute	Solubility in Cyclohexane	Solubility in Water	Organic or Inorganic?
NaCl			
Toluene			

C. Combustion

Compound	Flammable (Color of Flame)	Not Flammable	Organic or Inorganic?
NaCl			
Cyclohexane			

Laboratory D3

Report Sheet - Lab D3

From your observations of the chemical and physical properties of alkanes as organic compounds, complete the following table:

Property	Organic Compounds	Inorganic Compounds
Elements		
Bonding		
Melting points		
Strong odors		
Flammability		
Solubility		

Questions and Problems

Q.1 Describe three properties you can use to distinguish between organic and inorganic compounds.

Q.2 A white solid has no odor, is soluble in water, and is not flammable. Would you expect it to be an organic or an inorganic substance? Why?

Q.3 A clear liquid with a gasoline-like odor forms a layer when added to water. Would you expect it to be an organic or an inorganic substance? Why?

Report Sheet - Lab D3

D. Structures of Alkanes

D.1 Structure of methane		
Tetrahedral shape	Complete structural formula	Condensed structural formula

D.2 Structure of ethane	
Complete structural formula	Condensed structural formula

D.3 Structure of propane	
Complete structural formula	Condensed structural formula

Questions and Problems

Q.4 Write the correct name of the following alkanes:

a. $CH_3CH_2CH_3$ _____

b. $CH_3CH_2\underset{\underset{CH_3}{|}}{\overset{\overset{CH_3}{|}}{C}H}CHCH_3$ _____

c. $CH_3-\underset{\underset{}{}}{\overset{\overset{CH_3}{|}}{C}H}-\underset{\underset{}{}}{\overset{\overset{CH_3}{|}}{C}H}-CH_3$ _____

Laboratory D3

Report Sheet - Lab D3

Q.5 Write the condensed formulas for the following:

a. hexane   b. 2,3-dimethylpentane

E. Constitutional Isomers

E.1 Butane C$_4$H$_{10}$	
Complete structural formula	Condensed structural formula

E.2 2-Methylpropane	
Complete structural formula	Condensed structural formula

E.3 Physical Properties of Isomers of C$_4$H$_{10}$

Isomer	Molar Mass	Melting Point	Boiling Point	Density
Butane				
2-Methylpropane (isobutane)				

Questions and Problems

Q.6 In E.3, what physical property is identical for the two isomers of C$_4$H$_{10}$?

Properties and Structures of Alkanes

Report Sheet - Lab D3

Q.7   What physical properties are different for the isomers? Explain.

E.4 Isomers of C_5H_{12}

Complete structural formula	Condensed structural formula

Name:

Complete structural formula	Condensed structural formula

Name:

Complete structural formula	Condensed structural formula

Name:

E.5 Physical Properties of Isomers of C_5H_{12}

Isomer	Molar Mass	Melting Point	Boiling Point	Density
Pentane				
2-Methylbutane				
2,2-Dimethylpropane				

Laboratory D3

Report Sheet - Lab D3

Questions and Problems

Q.8 Write the condensed formulas for the five isomers of C_6H_{14}.

F. Cycloalkanes

Complete Structural Formula	Condensed Structural Formula	Geometric Formula
F.1 Three carbon atoms		
Name:		
F.2 Four carbon atoms		
Name:		
F.3 Five carbon atoms		

Measurement and Significant Figures

Goals
- Identify metric units used in measurement such as gram, meter, centimeter, millimeter, and milliliter.
- Correctly read a meter stick, a balance, and a graduated cylinder.
- State the correct number of significant figures in a measurement.

Text References

Review the following before performing this laboratory.

Timberlake; *General, Organic, and Biological Chemistry: Structures of Life*, Third Edition
Sections 1.1 to 1.5; Pages 14 to 32

or

Timberlake; *Chemistry: An Introduction to General, Organic, and Biological Chemistry*, Tenth Edition
Section 1.3; Pages 21 to 23

Discussion

Scientists and allied health personnel carry out laboratory procedures, take measurements, and report results accurately and clearly. How well they do these things can mean life or death to a patient. The system of measurement used in science, hospitals, and clinics is the metric system. The metric system is a *decimal system* in which measurements of each type are related by factors of 10. You use a decimal system when you change U.S. money. For example, 1 dime is the same as 10 cents or one cent is 1/10 of a dime. A dime and a cent are related by a factor of 10.

The metric system has one standard unit for each type of measurement. For example, the metric unit of length is the meter, whereas the U.S. system of measurement uses many units of length such as inch, foot, yard, and mile. Most of the rest of the world uses the metric system only. The most common metric units are listed in Table 1.1. SI (International Units) have the same 5 fundamental units as Metric Units plus units for electric current (ampere, A) and luminous intensity (candela, cd).

Table 1.1 *Metric Units*

Measurement	Metric Unit	Symbol
Length	meter	m
Mass	gram	g
Volume	liter	L
Temperature	degrees Celsius; Kelvin	°C; K
Time	second	s

Laboratory 1

A unit must always be included when reporting a measurement. For example, 5.0 m indicates a quantity of 5.0 meters in this measurement of length. Without the unit, we would not know the system of measurement used to obtain the number 5.0. It could have been 5.0 feet, 5.0 kilometers, or 5.0 inches. Thus, a unit is required to complete the measurement reported. For larger and smaller measurements, prefixes are attached in front of the standard unit. Some prefixes such as *kilo* are used for larger quantities; other prefixes such as *milli* are used for smaller quantities. The most common prefixes are listed in Table 1.2.

Table 1.2 *Some Prefixes in the Metric System*

Prefix	Symbol	Meaning
kilo	k	1000
deci	d	0.1 (1/10)
centi	c	0.01 (1/100)
milli	m	0.001 (1/1000)

Measured and Exact Numbers

When we measure the length, volume, or mass of an object, the numbers we report are called *measured numbers*. Suppose you got on a scale this morning and saw that you weighed 145 lb. The scale is a measuring tool and your weight is a measured number. Each time we use a measuring tool to determine a quantity, the result is a measured number.

Exact numbers are obtained when we count objects. Suppose you counted 5 beakers in your laboratory drawer. The number 5 is an exact number. You did not use a measuring tool to obtain the number. Exact numbers are also found in the numbers that define a relationship between two metric units or between two U.S. units. For example, the numbers in the following definitions are exact: 1 meter is equal to 100 cm; 1 foot has 12 inches. See Sample Problem 1.1.

Sample Problem 1.1

Describe each of the following as a measured or exact number:

a. 14 inches   b. 14 pencils   c. 60 minutes in 1 hour   d. 7.5 kg

Solution:

a. measured   b. exact   c. exact (definition)   d. measured

Significant Figures in Measurements

In measured numbers, all the reported figures are called *significant figures*. The first significant figure is the first nonzero digit. The last significant figure is always the estimated digit. Zeros between other digits or to the right of the decimal point in a decimal number are counted as significant figures. However, leading zeros are *not significant;* they are placeholders. Zeros are *not significant* in large numbers with no decimal points; they are placeholders needed to express the magnitude of the number.

When a number is written in scientific notation, all the figures in the coefficient are significant. Examples of counting significant figures in measured numbers are in Table 1.3 and Sample Problem 1.2.

Measurement and Significant Figures

Table 1.3 *Examples of Counting Significant Figures*

Measurement	Number of Significant Figures	Reason
455.2 cm	4	All nonzero digits are significant.
0.80 m	2	A following zero in a decimal number is significant.
50.2 L	3	A zero between nonzero digits is significant.
0.0005 lb	1	Leading zeros are not significant.
25,000 ft	2	Placeholder zeros are not significant.

Sample Problem 1.2

State the number of significant figures in each of the following measured numbers:

a. 0.00580 m   b. 132.08 g

Solution:

a. Three significant figures. The zeros after the decimal point are placeholder zeros, but the zero following nonzero digits is *significant*.
b. Five significant figures. The zero between nonzero digits is significant.

When you use a meterstick, or read the volume in a graduated cylinder, the measurement must be reported as precisely as possible. See Figure 1.1. The number of *significant figures* you can report depends on the lines marked on the measuring tool you use. For example, on a 50-mL graduated cylinder, the small lines represent a 1-mL volume. If the liquid level is between 21 mL and 22 mL, you know you can report 21 mL for certain. However, you can add one more digit (*the last digit*) to your reported value by estimating between the 1-mL lines. For example, if the volume level were halfway between the 21-mL and 22-mL lines, you would report the volume as 21.5 mL. If the volume level *is exactly* on the 21-mL line, you indicate this precision by adding a *significant zero* to give a measured volume of 21.0 mL. See Figure 1.2.

A.   Measuring Length

The standard unit of length in the metric system is the *meter* (*m*). Using an appropriate prefix, you can indicate a length that is greater or less than a meter as listed in Table 1.4. Kilometers are used in most countries for measuring the distance between two cities, whereas centimeters or millimeters are used for small lengths.

Table 1.4 *Some Metric Units Used to Measure Length*

Length	Symbol	Meaning
1 kilometer	km	1000 meters (m) or 10^3 m
1 decimeter	dm	0.1 m (1/10 m) or 10^{-1} m
1 centimeter	cm	0.01 m (1/100 m) or 10^{-2} m
1 millimeter	mm	0.001 m (1/1000 m) or 10^{-3} m

Laboratory 1

Figure 1.1 A meterstick divided into centimeters (cm)

A *meterstick* is divided into 100 cm as seen in Figure 1.1. The smallest lines are centimeters. That means that each measurement you make can be certain to the centimeter. The final digit in a precise measurement is obtained by estimating. For example, the shorter line in Figure 1.1 reaches the 44-cm mark and is about halfway to 45 cm. We report its length as 44.5 cm. The last digit (0.5) is the estimated digit. If the line appears to end at a centimeter mark, then the estimated digit is 0.0 cm. The longer line in Figure 1.1 appears to end right at the 67-cm line, which is indicated by reporting its length as 67.0 cm.

B.   Measuring Volume

The volume of a substance measures the space it occupies. In the metric system, the unit for volume is the *liter* (*L*). Prefixes are used to express smaller volumes such as deciliters (dL) or milliliters (mL). One cubic centimeter (cm^3 or cc) is equal to 1 mL. The terms mL and cc are used interchangeably. See Table 1.5.

Table 1.5 *Some Metric Units Used to Measure Volume*

Unit of Volume	Symbol	Meaning
1 kiloliter	kL	1000 liters (L) or 10^3 L
1 deciliter	dL	0.1 L (1/10 L) or 10^{-1} L
1 milliliter	mL	0.001 L (1/1000 L) or 10^{-3} L

In the laboratory, the volume of a liquid can be measured in a graduated cylinder (Figure 1.2). Set the cylinder on a level surface and bring your eyes even with the liquid level. Notice that the water level is not a straight line but curves downward in the center. This curve, called a *meniscus,* is read at its lowest point (center) to obtain the correct volume measurement for the liquid. Your eyes should be aligned with the bottom of the meniscus in order to avoid error in making the reading. In this graduated cylinder, the volume of the liquid is 42.0 mL.

Measurement and Significant Figures

Figure 1.2 Reading a volume of 42.0 mL in a graduated cylinder

On large cylinders, the lines may represent volumes of 2 mL, 5 mL, or 10 mL. On a 250-mL cylinder, the marked lines usually represent 5 mL. On a 1000-mL cylinder, each line may be 10 mL. Then your precision on a measurement will be to the milliliter or mL.

C. Measuring Mass

The *mass* of an object indicates the amount of matter present in that object. The *weight* of an object is a measure of the attraction that Earth has for that object. Because this attraction is proportional to the mass of the object, we will use the terms *mass* and *weight* interchangeably even through the two terms are not identical.

In the metric system, the unit of mass is the *gram (g)*. A larger unit, the *kilogram (kg)*, is used in measuring a patient's weight in a hospital. A smaller unit of mass, the *milligram (mg)*, is often used in the laboratory. See Table 1.6.

Table 1.6 *Some Metric Units Used to Measure Mass*

Mass	Symbol	Meaning
kilogram	kg	1000 g or 10^3 g
gram	g	1000 mg or 10^3 mg
milligram	mg	1/1000 g (0.001 g) or 10^{-3} g

Lab Information

Time:	2 hr
Comments:	Tear out the report sheets and place them beside the experimental procedures as you work.
	Determine the markings on each measuring tool before you measure.
	Record all the possible numbers for a measurement including the estimated digit.
	Write a unit of measurement after each measured number.
Related Topics:	Significant figures, measured and exact numbers, metric prefixes

Experimental Procedures   *GOGGLES REQUIRED!*

Laboratory 1

A. Measuring Length

Materials: Meterstick, string

Section	Step	Procedure	Check when finished
A.1	1	Observe the marked lines on a meterstick. Identify the lines that represent centimeters and millimeters. Determine how you will estimate between the smallest lines.	

Section	Step	Procedure	Check when finished
A.2	1	Use the meterstick to make the length measurements (cm) indicated on the report sheet. String may be used to determine the distance around your wrist. Include the estimated digit in each measurement.	

Section	Step	Procedure	Check when finished
A.3	1	Indicate the estimated digit and the number of significant figures in each measurement in A.2. See Sample Problem 1.3.	

Sample Problem 1.3

What is the estimated digit in each of the following measured masses?

a. beaker 42.18 g   b. pencil 11.6 g

Solution:

a. hundredths place (0.08 g)   b. tenths place (0.6 g)

Section	Step	Procedure	Check when finished
A.4	1	Measure the length of the line on the report sheet at A.4.	

B. Measuring Volume

Materials: Display of graduated cylinders with liquids, 50-mL, 100-mL, 250-mL, and 500-mL (or larger) graduated cylinders, test tube, solid object

Measurement and Significant Figures

Section	Step	Procedure	Check when finished
B.1	1	**Volume of a liquid:** Determine the volumes of the liquids in a display of graduated cylinders. Be as precise as you can. For example, each line marked on a 50-mL graduated cylinder measures 1 mL. By estimating the volume *between* the 1-mL markings, you can report a volume to a tenth (0.1) of a milliliter. Indicate the estimated digit and the number of significant figures in each measurement.	

Section	Step	Procedure	Check when finished
B.2	1	**Volume of a test tube:** Fill a small test tube to the rim.	
B.2	2	Carefully pour the water into a small graduated cylinder.	
B.2	3	State the volume represented by the smallest marked lines on the cylinder.	
B.2	4	Record the volume of the water and state the estimated digit.	
B.2	5	Fill the test tube again and pour the water into a medium-sized graduated cylinder.	
B.2	6	Record.	
B.2	7	Repeat this process using a large graduated cylinder.	
B.2	8	Record.	

Section	Step	Procedure	Check when finished
B.3	1	**Volume of a solid by volume displacement:** When an object is submerged in water, it displaces its own volume of water, causing the water level to rise. The volume of the object is the difference in the water level before and after the object is submerged. See Figure 1.3.	
B.3	2	Obtain a graduated cylinder that will hold the solid.	
B.3	3	Place water in the graduated cylinder until it is about half full.	
B.3	4	Carefully record the volume of water.	
B.3	5	Tie a piece of thread around the heavy solid object.	
B.3	6	Slowly submerge the solid under the water.	
B.3	7	Record the new volume of the water.	
B.3	8	Calculate the volume (mL) displaced by the solid.	

Laboratory 1

Figure 1.3 Using volume displacement to determine the volume of a solid

C. Measuring Mass

Materials: Balance, objects to weigh (beaker, rubber stopper, evaporating dish), unknown mass

Section	Step	Procedure	Check when finished
C.1	1	**Have your instructor show you how to use a laboratory balance.**	
C.1	2	Determine the mass of the listed objects from your lab drawer. If you are using a triple beam balance, be sure that all of your recorded measurements include an estimated digit.	

Section	Step	Procedure	Check when finished
C.2	1	Obtain an object of unknown mass from your instructor.	
C.2	2	Record the code number.	
C.2	3	Determine its weight.	
C.2	5	Record.	
C.2	6	Check your result with the instructor.	

Report Sheet - Lab 1

Date _____   Name _____
Section _____   Team _____
Instructor _____   _____

Pre-Lab Study Questions

1. What are the standard units of length, mass, volume, and temperature in the metric (SI) system?

2. Why is the metric (SI) system called a decimal system of measurement?

3. What is the purpose of using prefixes in the metric (SI) system?

Laboratory 1

Report Sheet - Lab 1

4. Give the name or the abbreviation for each metric unit listed. State the property that it measures.

Name of Unit	Abbreviation	Property Measured
_____	L	_____
centimeter	_____	_____
_____	km	_____
_____	mg	_____

5. Describe each of the following as a measured number or an exact number:

5 books _____   12 roses _____

5 lb _____   12 inches in 1 foot _____

9.25 g _____   361 miles _____

0.035 kg _____   100 cm in 1 m _____

6. How do you determine the last digit in any measured number?

7. If several different students measured the same length with a meterstick, would they get exactly the same result?

10

Measurement and Significant Figures

Report Sheet - Lab 1

Date _____ Name _____
Section _____ Team _____
Instructor _____   _____

A. Measuring Length

A.1 What units are represented by the numbers marked on the meterstick? _____
     What do the small lines marked on the meterstick represent? _____
     Complete the following statements:
     There are _____ centimeters (cm) in 1 meter (m).
     There are _____ millimeters (mm) in 1 meter (m).
     There are _____ millimeters (mm) in 1 centimeter (cm).

A.2

Item	Length	A.3 Estimated Digit	Number of Significant Figures
Width of little fingernail	_____	_____	_____
Distance around your wrist	_____	_____	_____
Length of your shoe	_____	_____	_____
Your height	_____	_____	_____

A.4 Length of line

_____

Your measurement   _____
Other students' values _____ _____
                        _____ _____

Questions and Problems

Q.1 What digits in the measurements for the line by other students are the same as yours and which are different?

Q.2 Why could the measured values obtained by other students be different from yours?

Laboratory 1

Report Sheet - Lab 1

B. Measuring Volume

B.1   **Volume of a liquid** (*Include units for every measurement*)

	Cylinder 1	Cylinder 2	Cylinder 3
Volume (mL)			

B.2   **Volume of a test tube**

	Small Cylinder _____ mL	Medium Cylinder _____ mL	Large Cylinder _____ mL
Volume represented by smallest line			
Volume (mL) of water in test tube			
Estimated digit			

Which cylinder is the best choice for measuring the volume of water in a test tube? Explain.

B.3   **Volume of a solid by volume displacement**

Initial volume of water   _____

Volume of water and submerged solid   _____

Volume of solid   _____

C. Measuring Mass

	Item	Mass	Number of Significant Figures
C.1	Beaker	_____	_____
	Stopper	_____	_____
	Evaporating dish	_____	_____
C.2	Unknown #____ _____		_____

Questions and Problems

Q.3   State the number of significant figures in each of the following measurements:

   4.5 m   _____   204.52 g   _____

   0.0004 L   _____   625,000 mm   _____

   805 lb   _____   34.80 km   _____

Q.4   Indicate the estimated digit in each of the following measurements:

   1.5 cm   _____   4500 mi   _____

   0.0782 m   _____   42.50 g   _____

2
Density and Specific Gravity

Goals
- Calculate the density of a substance from measurements of its mass and volume.
- Calculate the specific gravity of a liquid from its density.
- Determine the specific gravity of a liquid using a hydrometer.

Text References

Review the following before performing this laboratory.

Timberlake; *General, Organic, and Biological Chemistry: Structures of Life*, Third Edition
Section 1.8; Pages 43 to 49

or

Timberlake; *Chemistry: An Introduction to General, Organic, and Biological Chemistry*, Tenth Edition
Section 1.8; Pages 40 to 43

Discussion

A. Density of a Solid

To determine the density of a substance, you need to measure both its mass and its volume. You have carried out both of these procedures in previous labs. From the mass and volume, the density is calculated. If the mass is measured in grams and the volume in milliliters, the density will have the units of g/mL.

$$\text{Density of a substance} = \frac{\text{Mass of substance}}{\text{Volume of substance}} = \frac{\text{g of substance}}{\text{mL of substance}}$$

B. Density of a Liquid

To determine the density of a liquid, you need the mass and volume of the liquid. The mass of a liquid is determined by weighing. The mass of a container is obtained and then a certain volume of liquid is added and the combined mass determined. Subtracting the mass of the container gives the mass of the liquid. From the mass and volume, the density is calculated.

$$\text{Density of liquid} = \frac{\text{Mass (g) of liquid}}{\text{Volume (mL) of liquid}}$$

Laboratory 2

C. Specific Gravity

The specific gravity of a liquid is a comparison of the density of that liquid with the density of water, which is 1.00 g/mL at 4°C.

$$\text{Specific gravity} = \frac{\text{Density of a substance (g/mL)}}{\text{Density of water (1.00 g/mL)}}$$

Specific gravity is a number with no units; the units of density (g/mL) have canceled out. This is one of the few measurements in chemistry written without any units.

Using a hydrometer The specific gravity of a fluid can be determined by using a hydrometer. Small hydrometers (urinometers) are used in the hospital to determine the specific gravity of urine. Another type of hydrometer is used to measure the specific gravity of the fluid in your car battery. A hydrometer placed in a liquid is spun slowly to keep it from sticking to the sides of the container. The scale on the hydrometer is read at the lowest (center) point of the meniscus of the fluid. Read the specific gravity on the hydrometer to 0.001. See Figure 2.1.

Figure 2.1 Measuring specific gravity using a hydrometer

Density and Specific Gravity

Lab Information

Time: 3 hr

Comments: Tear out the report sheets and place them beside the procedures.

Round off the calculator answers to the correct number of significant figures.

Dispose of liquids properly as directed by your instructor.

Related Topics: Mass, volume, prefixes, significant figures, density, specific gravity

Experimental Procedures *GOGGLES REQUIRED!*

A. Density of a Solid

Materials: Metal object, string or thread, graduated cylinder

Mass of a solid

Section	Step	Procedure	Check when finished
A.1	1	Obtain a solid metal object.	
A.1	2	Weight it and record.	

Volume of the solid by displacement

Section	Step	Procedure	Check when finished
A.2	1	Obtain a graduated cylinder that is large enough to hold the solid metal object.	
A.2	2	Add water until the cylinder is about half full.	
A.2	3	Read the water level carefully and record.	
A.2	4	If the solid object is heavy, lower it into the water by attaching a string or thread.	
A.2	5	While the solid object is submerged in the water, record the final water level.	
A.2	6	Calculate the volume of the solid. Volume of solid = Final water level – initial water level	

Laboratory 2

Calculating the density of the solid

Section	Step	Procedure	Check when finished
A.3	1	Calculate the density (g/mL) of the solid by dividing its mass (g) by its volume (mL). Be sure to determine the correct number of significant figures in your calculated density value. (See equation below.)	

Identifying the solid

Section	Step	Procedure	Check when finished
A.4	1	If your instructor indicates that the solid is made of one of the substances in Table 2.1, use the density you calculated in A.3 to identify the metal from the known values for density.	

$$\text{Density of solid} = \frac{\text{Mass (g) of solid}}{\text{Volume (mL) of solid}}$$

Table 2.1 *Density Values of Some Metals*

Substance	Density (g/mL)
Aluminum	2.7
Brass	8.4
Copper	8.9
Iron	7.9
Lead	11.3
Nickel	8.9
Tin	7.3
Zinc	7.1

B. Density of a Liquid

Materials: 50-mL graduated cylinder, two liquid samples, 100-mL or 250-mL beaker

Volume of a liquid

Section	Step	Procedure	Check when finished
B.1	1	Place about 20 mL of water in a 50-mL graduated cylinder.	
B.1	2	Record. *(Do not use the markings on beakers to measure volume; they are not precise.)*	

Density and Specific Gravity

Mass of a liquid

Section	Step	Procedure	Check when finished
B.2	1	The mass of a liquid is found by weighing by difference. Weigh a small dry beaker. *Taring a container on an electronic balance:* The mass of a container on an electronic balance can be set to 0 by pressing the *tare* bar. As a substance is added to the container, the mass shown on the readout is for the *substance* only. (When a container is *tared*, it is not necessary to subtract the mass of the beaker.)	
B.2	2	Record the weight.	
B.2	3	Pour the liquid into the beaker.	
B.2	4	Weigh the beaker and then the beaker and the liquid.	
B.2	5	Record the weight.	
B.2	6	Calculate the mass of the liquid.	

Density of a liquid

Section	Step	Procedure	Check when finished
B.3	1	Calculate the density of the liquid by dividing its mass (g) by the volume (mL) of the liquid.	
B.3	2	Repeat B.2 step 1 to step 6 and B.3 step 1 for all of the liquids provided in the laboratory.	

$$\text{Density of liquid} = \frac{\text{Mass (g) of liquid}}{\text{Volume (mL) of liquid}}$$

C. Specific Gravity

Materials: Water, liquids used in part B in graduated cylinders with hydrometers

Section	Step	Procedure	Check when finished
C.1	1	Calculate the specific gravity (sp gr) of each liquid you used in B by dividing its density by the standard density of water (1.00 g/mL). See the equation below.	

Laboratory 2

Section	Step	Procedure	Check when finished
C.2	1	Read the hydrometer set in a graduated cylinder containing the same liquid you used in the density section.	
C.2	2	Record. (Some hydrometers use the European decimal point, which is a comma. The value 1,000 on a European scale is read as 1.000. Record specific gravity as a decimal number.)	

$$\text{Specific gravity} = \frac{\text{Density of a substance (g/mL)}}{\text{Density of water (1.00 g/mL)}}$$

Density and Specific Gravity

Report Sheet - Lab 2

Date _____ Name _____

Section _____ Team _____

Instructor _____ _____

Pre-Lab Study Questions

1. What property of oil makes it float on water?

2. Why would heating the gas in an air balloon make the balloon rise?

3. What is the difference between density and specific gravity?

Laboratory 2

Density and Specific Gravity

Report Sheet - Lab 2

Date _____   Name _____

Section _____   Team _____

Instructor _____   _____

A.   Density of a Solid

A.1   **Mass of the solid**   _____

A.2   **Volume of the solid by displacement**

     Initial water level (mL)   _____

     Final water level with solid (mL)   _____

     Volume of solid (mL)   _____

A.3   **Calculate the density of the solid**   _____ g/mL

     (Show calculations.)

A.4   Type of metal   _____

Questions and Problems *(Show complete setups.)*

Q.1   An object made of aluminum has a mass of 8.37 g. When it was placed in a graduated cylinder containing 20.0 mL of water, the water level rose to 23.1 mL. Calculate the density and specific gravity of the object.

Laboratory 2

Report Sheet - Lab 2

B. Density of a Liquid

		Liquid 1	Liquid 2
B.1	**Volume of liquid**		
	Type of liquid	_____	_____
	Volume (mL)	_____	_____
B.2	**Mass of liquid**		
	Mass of beaker	_____	_____
	Mass of beaker + liquid	_____	_____
	Mass of liquid	_____	_____
B.3	**Density of liquid**		
	Density	_____	_____

(Show calculations for density.)

C. Specific Gravity

C.1	Specific gravity	_____	_____

(Calculated using B.3. Show calculations.)

C.2	Specific gravity		
	(Hydrometer reading)	_____	_____

Density and Specific Gravity

Report Sheet - Lab 2

How does the *calculated* specific gravity compare to the hydrometer reading for each liquid?

Questions and Problems *(Show complete setups.)*

Q.2 What is the mass of a solution that has a density of 0.775 g/mL and a volume of 50.0 mL?

Q.3 What is the volume of a solution that has a specific gravity of 1.2 and a mass of 185 g?

Laboratory 2

3

Atomic Structure and Electronic Arrangement

Goals

- Write the correct symbols or names of some elements.
- Describe some physical properties of the elements you observe.
- Categorize an element as a metal or nonmetal from its physical properties.
- Given the complete symbol of an atom, determine its mass number, atomic number, and the number of protons, neutrons, and electrons.
- Describe the color of a flame produced by an element.
- Use the color of a flame to identify an element.
- Draw models of atoms.

Text References

Review the following before performing this laboratory.

Timberlake; *General, Organic, and Biological Chemistry: Structures of Life*, Third Edition
Sections 3.1 to 3.8; Pages 95 to 131

or

Timberlake; *Chemistry: An Introduction to General, Organic, and Biological Chemistry*, Tenth Edition
Sections 3.1 to 3.8; Pages 83 to 111

Discussion

Primary substances, called elements, build all the materials about you. Some look similar, but others look unlike anything else. In this experiment, you will describe the physical properties of elements in a laboratory display and determine the location of elements on a blank periodic table.

A. Physical Properties of Elements

Metals are elements that are usually shiny or have a metallic luster. They are usually good conductors of heat and electricity, ductile (can be drawn into a wire), and malleable (can be molded into a shape). Some metals such as sodium or calcium may have a white coating of oxide formed by reacting with oxygen in the air. If these are cut, you can see the fresh shiny metal underneath. In contrast, nonmetals are not good conductors of heat and electricity, are brittle (not ductile), and appear dull, not shiny.

Laboratory 3

B. Periodic Table

The periodic table, shown on the inside front cover of this lab manual and your textbook, contains information about each of the elements. On the table, the horizontal rows are *periods,* and the vertical columns are *groups.* Each group contains elements that have similar physical and chemical properties. The groups are numbered across the top of the chart. Elements in Group 1A are the *alkali metals,* elements in Group 2A are the *alkaline earths,* and Group 7A contains the *halogens.* Group 8A contains the *noble gases,* which are elements that are not very reactive compared to other elements. A dark zigzag line that looks like a staircase separates the *metals* on the left side from the *nonmetals* on the right side.

C. Subatomic Particles

There are different kinds of atoms for each of the elements. Atoms are made up of smaller bits of matter called *subatomic particles. Protons* are positively charged particles, *electrons* are negatively charged, and *neutrons* are neutral (no charge). In an atom, the protons and neutrons are tightly packed in the tiny center called the *nucleus.* Most of the atom is empty space, which contains fast-moving electrons. Electrons are so small that their mass is considered to be negligible compared to the mass of the proton or neutron. The *atomic number* is equal to the number of protons. The *mass number* of an atom is the number of protons plus the number of neutrons.

atomic number = number of protons (p^+)

mass number = sum of the number of protons and neutrons ($p^+ + n^0$)

D. Isotopes

Isotopes are atoms of the same element that differ in the number of neutrons. This means that isotopes of an element have the same number of protons, but different mass numbers. The following example represents the symbol of a sulfur isotope that has 16 protons and 18 neutrons.

Complete Symbol of an Isotope	Meaning
mass number (p^+ and n^0) → 34	This atom has 16 protons and 18 neutrons.
symbol of element → S	The element is sulfur.
atomic number (p^+) → 16	The atom has 16 protons.

Atomic Structure and Electronic Arrangement

E. Flame Tests

The chemistry of an element strongly depends on the arrangement of the electrons. The energy levels for electrons of atoms of the *first 20 elements* have the following number of electrons.

Electron Arrangement for Elements 1–20

Level 1 ($2e^-$)   Level 2 ($8e^-$)   Level 3 ($8e^-$)   Level 4 ($2e^-$)

When electrons absorb specific amounts of energy, they can attain higher energy levels. In order to return to the lower, more stable energy levels, electrons release energy. If the energy released corresponds to an energy of visible light, the emission produces a color that can be detected by the human eye.

When heated, many of the elements in Groups 1A and 2A produce colorful flames. Each element produces a characteristic color. When the light from one of these flames passes through a glass prism or crystal, a series of colored lines appears. The spaces between lines appear dark. Such a series of lines, known as a *line spectrum,* is used to identify elements in water, food, the Sun, stars, and on other planets.

F. Drawing Models of Atoms

A model of an atom can be drawn to show the number of proteins and neutrons in the nucleus with the electrons shown in energy shells. This is an oversimplification of the true nature of atoms, but the model does illustrate the relationship of the subatomic particles. By observing the electron arrangement, you can determine the number of valence elections for the atoms of that element.

We can illustrate the model of an atom of boron with a mass number of 11. With an atomic number of 5, boron has 5 protons. To find the number of neutrons, the atomic number (number of protons) is subtracted from mass number of 11 (for this isotope) to give 6 neutrons. Boron has 3 valence electrons, which puts boron in Group 3. See Figure 3.1.

Figure 3.1 Model of Boron atom

Laboratory 3

Lab Information

Time: 3 hr

Comments: Obtain a periodic table or use the one on the inside cover of your textbook.

Tear out the Lab 5 report sheets and place them beside the procedures.

In neutral atoms, the number of electrons is equal to the number of protons.

Related Topics: Electrons and protons, energy levels, and electron arrangement

Experimental Procedures   *GOGGLES REQUIRED!*

A. Physical Properties of Elements

Materials: A display of elements

Section	Step	Procedure	Check when finished
A.1	1	Observe the elements in the laboratory display of elements.	
A.1	2	In the report sheet, write the symbol and atomic number for each element listed.	
A.1	3	Describe some physical properties such as color and luster.	
A.1	4	From your observations, identify each element as a metal (M) or a nonmetal (NM).	

B. Periodic Table

Materials: Periodic table, colored pencils, display of elements

Section	Step	Procedure	Check when finished
B.1	1	On the incomplete periodic table provided in the report sheet, write the atomic numbers and symbols of the elements you observed in part A.	
B.1	2	Write the period numbers for each of the horizontal rows shown.	
B.1	3	Using different colors, shade in the columns that contain the alkali metals, alkaline earths, halogens, and noble gases.	
B.1	4	With another color, shade in the transition elements. Draw a heavy line to separate the metals and nonmetals.	

Section	Step	Procedure	Check when finished

Atomic Structure and Electronic Arrangement

Section	Step	Procedure	Check when finished
B.2	1	Without looking at the display of elements, use the periodic table to decide whether the elements listed on the report sheet would be metals or nonmetals; shiny or dull.	
B.2	2	After you complete your predictions, observe those same elements in the display to see if you predicted correctly.	

C. Subatomic Particles

Section	Step	Procedure	Check when finished
C.1	1	For each of the neutral atoms described in the table, write the atomic number, mass number, and number of protons, neutrons, and electrons.	

D. Isotopes

Section	Step	Procedure	Check when finished
D.1	1	Complete the information for each of the isotopes of calcium: the complete nuclear symbol and the number of protons, neutrons, and electrons.	

E. Flame Tests

Materials: Bunsen burner, spot plate, flame-test (nichrome) wire, cork, 100-mL beaker, 1 M HCl, 0.1 M solutions (dropper bottles): $CaCl_2$, KCl, $BaCl_2$, $SrCl_2$, $CuCl_2$, NaCl, and unknown solutions

Section	Step	Procedure	Check when finished
E.1	1	Obtain a spot plate, flame-test wire, and cork stopper.	
E.1	2	Bend one end of the flame-test wire into a small loop and secure the other end in a cork stopper.	
E.1	3	Pour a small amount of 1 M HCl into a 100-mL beaker. **CAUTION 1 M HCl is corrosive! Be careful when you use it. Wash off any HCl spills on the skin with tap water for 10 minutes.**	
E.1	4	Rinse the spot plate in distilled water.	
E.1	5	Place 6–8 drops of each test solution in separate indentations of the spot plate.	

Laboratory 3

Section	Step	Procedure	Check when finished
E.1	6	Label the spot plate diagram in the laboratory report to match the solutions. Be careful not to mix the different solutions.	
E.1	7	Adjust the flame of a Bunsen burner until it is nearly colorless.	
E.1	8	Clean the test wire by dipping the loop in the HCl in the beaker and placing it in the flame of the Bunsen burner. (If you see a strong color in the flame while heating the wire, dip it in the HCl again. Repeat until the color is gone.)	

Section	Step	Procedure	Check when finished
E.2	1	**Observing Flame Colors:** Dip the cleaned wire in one of the solutions on the spot plate. Make sure that a thin film of the solution adheres to the loop. See Figure 3.2.	
E.2	2	Move the loop of the wire into the lower portion of the flame and record the color you observe.	
E.2	3	For each solution, it is the first element in the formula that is responsible for color.	
E.2	4	Repeat each flame test until you can describe the color of the flame produced.	
E.2	5	Clean the wire and repeat the flame test with the other solutions.	

Section	Step	Procedure	Check when finished
E.3	1	**Identifying solutions:** Obtain unknown solutions as indicated by your instructor and record their code letters.	
E.3	2	Place 6 to 8 drops of each unknown solution in a clean spot plate.	
E.3	3	Use the flame-test procedure to determine the identity of the unknown solution. You may wish to recheck the flame color of the known solution that best matches the flame color of an unknown. For example, if you think your unknown is KCl, recheck the color of the KCl solution to confirm.	

Figure 3.2 Using a flame-test wire to test for flame color

Note: The color of potassium in the KCl flame is short-lived. Be sure to observe the color of the flame from the KCl solution within the first few seconds of heating.

F. Drawing Models of Atoms

Draw a model of each atom listed on the laboratory report. Indicate the number of valence electrons and the group number for the element.

Atomic Structure and Electronic Arrangement

Report Sheet - Lab 3

Date _____ Name _____
Section _____ Team _____
Instructor _____   _____

Pre-Lab Study Questions

1. Describe the periodic table.

2. Where are the alkali metals and the halogens located on the periodic table?

3. On the following list of elements, circle the symbols of the transition elements and underline the symbols of the halogens:

   Mg   Cu   Br   Ag   Ni   Cl   Fe   F

Laboratory 3

4. Complete the list of names of elements and symbols:

Name of Element	Symbol	Name of Element	Symbol
Potassium			Na
Sulfur			P
Nitrogen			Fe
Magnesium			Cl
Copper			Ag

5. What is the color of a neon light?

6. Why does a sodium street lamp give off a different color light than a neon light?

7. What are the energy levels of electrons?

8. Why do some elements produce colorful flames?

9. How can you identify an unknown using a flame test?

34

Atomic Structure and Electronic Arrangement

Report Sheet - Lab 3

Date _____   Name _____
Section _____   Team _____
Instructor _____   _____

A.   Physical Properties of Elements

Element	Symbol	Atomic Number	Color	Luster	Metal/Nonmetal
Aluminum	_____	_____	_____	_____	_____
Carbon	_____	_____	_____	_____	_____
Copper	_____	_____	_____	_____	_____
Iron	_____	_____	_____	_____	_____
Magnesium	_____	_____	_____	_____	_____
Nickel	_____	_____	_____	_____	_____
Nitrogen	_____	_____	_____	_____	_____
Oxygen	_____	_____	_____	_____	_____
Phosphorus	_____	_____	_____	_____	_____
Silicon	_____	_____	_____	_____	_____
Silver	_____	_____	_____	_____	_____
Sulfur	_____	_____	_____	_____	_____
Tin	_____	_____	_____	_____	_____
Zinc	_____	_____	_____	_____	_____

Laboratory 3

Report Sheet - Lab 3

B. Periodic Table

B.1

Questions and Problems

Q.1 From their positions on the periodic table, categorize the following elements as metals (M) or nonmetals (NM).

Na _____   S _____   Cu _____   F _____   Fe _____   C _____

Ca _____

Q.2 Give the name of each of the following elements:

a. Noble gas in Period 2 _____   b. Halogen in Period 2 _____

c. Alkali metal in Period 3 _____   d. Halogen in Period 3 _____

e. Alkali metal in Period 4 _____

B.2

Element	Metal/Nonmetal	Prediction: Shiny or Dull	Correct? Yes/No
Chromium			
Gold			
Lead			
Cadmium			
Silicon			

Report Sheet - Lab 3

C. Subatomic Particles

Element	Atomic Number	Mass Number	Protons	Neutrons	Electrons
Iron				30	
		27			13
			19	20	
Bromine		80			
Gold		197			
			53	74	

D. Isotopes

Nuclear Symbol	Protons	Neutrons	Electrons
$^{40}_{20}Ca$			
	20	22	
$^{43}_{20}Ca$			
		24	20
$^{46}_{20}Ca$			

Questions and Problems

Q.3   A neutral atom has a mass number of 80 and has 45 neutrons. Write its complete symbol.

Laboratory 3

Report Sheet - Lab 3

Q.4 An atom has two more protons and two more electrons than the atom in question 3. What is its complete symbol?

D. Flame Tests

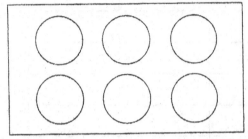
Spot plate diagram

Solution	Element	Color of Flame
CaCl$_2$	Ca	_____
KCl	K	_____
BaCl$_2$	Ba	_____
SrCl$_2$	Sr	_____
CuCl$_2$	Cu	_____
NaCl	Na	_____

Unknown Solution(s)

Identification letter   ☐   ☐   ☐

   Color of flame   _____   _____   _____

   Element present _____   _____   _____

Questions and Problems

Q.5 You are cooking spaghetti in water you have salted with NaCl. You notice that when the water boils over, it causes the flame of the gas burner to turn bright orange. How would you explain the appearance of a color in the flame?

Atomic Structure and Electronic Arrangement

F. Drawing Models of Atoms

Atom	Model of Atom	Number of Valence Electrons	Group Number
$^{7}_{3}\text{Li}$			
$^{14}_{7}\text{N}$			
$^{25}_{12}\text{Mg}$			
$^{27}_{13}\text{Al}$			
$^{37}_{17}\text{Cl}$			
$^{34}_{16}\text{S}$			

Questions and Problems

Q.6 Write the electron arrangement for the following elements:

Element	Energy Level			
	1	2	3	4
P				
Na				
F				
C				
Ca				

4

Nuclear Radiation

Goals
- Observe the use of a Geiger-Müeller radiation detection tube.
- Determine the effect of shielding materials, distance, and time of radiation.
- Complete a nuclear equation.

Text References

Review the following before performing this laboratory.

Timberlake; *General, Organic, and Biological Chemistry: Structures of Life*, Third Edition
Sections 4.1 to 4.3; Pages 139 to 152

or

Timberlake; *Chemistry: An Introduction to General, Organic, and Biological Chemistry*, Tenth Edition
Sections 9.1 to 9.3; Pages 318 to 333

Discussion

Radioactivity occurs when a proton or neutron breaks down in the nucleus of an unstable atom or the particles in the nucleus are rearranged. Then a particle or energy called *nuclear radiation* is emitted from the nucleus. The nucleus has undergone *nuclear decay*. The most typical kinds of radiation include alpha particles (α), beta particles (β), and gamma rays (γ).

alpha decay $\quad ^{147}_{62}Sm \rightarrow {}^{143}_{60}Nd + {}^{4}_{2}He \quad$ *alpha particle* (α)

beta decay $\quad ^{40}_{20}Ca \rightarrow {}^{40}_{21}Sc + {}^{0}_{-1}e \quad$ *beta particle* (β)

gamma decay $\quad ^{167}_{68}Er \rightarrow {}^{167}_{68}Er + {}^{0}_{0}\gamma \quad$ *gamma ray* (γ)

If radiation passes through the cells of the body, the cells may be damaged. You can protect yourself by using shielding materials, by limiting the amount of time near radiation sources, and by keeping a reasonable distance from the radioactive source.

To detect radiation, a device such as a Geiger-Müeller tube is used. Radiation passes through the gas held within the tube, producing ion pairs. These charged particles emit bursts of current that are converted to flashes of light and audible clicks. In this experiment your teacher will demonstrate the use of the Geiger-Müeller tube to test the effects of shielding, time, and distance.

Laboratory 4

Lab Information

Time: 2 hr

Comments: Tear out the report sheets and place them beside the procedures.
Follow your instructor's directions for protection from radiation.

Related Topics: Radioactivity, alpha particles, beta particles, gamma rays, shielding, nuclear decay, nuclear equations

Experimental Procedures GOGGLES REQUIRED!

(This experiment will be done as a demonstration.)

A. Background Count

Materials: Geiger-Müeller radiation detection tube

Section	Step	Procedure	Check when finished
A.1	1	The level of radiation that occurs naturally is called *background radiation*. Set the radiation counter at the proper voltage for operating level.	
A.1	2	Let it warm up for 5 minutes.	
A.1	3	Remove all sources of radiation near the counter.	
A.1	4	Count any radiation present in the room by operating the counter for 1 minute.	
A.1	5	Record the counts.	
A.1	6	Repeat the background count (steps 4 and 5) for two more 1-minute intervals.	

Section	Step	Procedure	Check when finished
A.2	1	Total the counts in A.1, and divide by 3. This value represents the average background level of radiation in counts per minute (cpm). This background count is the natural level of radiation that constantly surrounds and strikes us. For the rest of this experiment, this background radiation must be subtracted to give the radiation from the radioactive source alone.	

Nuclear Radiation

B. Radiation from Radioactive Sources

Materials: Geiger-Müeller radiation detection tube, meterstick, 3–4 radioactive sources of alpha- and beta-radiation samples to test for radioactivity: Fiestaware™, minerals, old lantern mantles containing thorium compounds, camera lenses, old watches with radium-painted numbers on dials, smoke detectors containing Am-241, antistatic devices for records and film containing Po-210, some foods such as salt substitute (KCl), cream of tartar, instant tea, instant coffee, dry seaweed

Section	Step	Procedure	Check when finished
B.1	1	Place one of the radioactive sources at a distance of 10–20 cm from the detection tube.	
B.1	2	Record the radiation emitted for 1 minute.	
B.1	3	Subtract the background count to obtain the radiation emitted by the source alone.	
B.1	4	Test other sources for 1 minute at the same distance.	

C. Effect of Shielding, Time, and Distance

Materials: Geiger-Müeller radiation detection tube, meterstick, one of the radioactive sources used in part B, shielding materials: lead, paper, glass, cardboard, etc.

Section	Step	Procedure	Check when finished
C.1	1	Place the radiation source at the same distance from the detection tube as it was in part B.	
C.1	2	Place various shielding materials (cardboard, paper, several pieces of cardboard, lead sheet, etc.) between the radioactive source and the detection tube. See Figure 4.1. Do not vary the distance of the source from the tube.	
C.1	3	Record the type of shielding used.	
C.1	4	For each type of shielding, record the radiation in counts per minute obtained from the source for 1 minute.	
C.1	5	Subtract the background count from each result to find the amount of radiation in cpm allowed by each type of shielding.	
C.1	6	Repeat steps 2 through 5 for the different types of shielding.	

Laboratory 4

Figure 4.1 Measuring radiation using shielding

Section	Step	Procedure	Check when finished
C.2	1	The more time you spend near a radioactive source, the greater the amount of radiation you receive. Keeping the distance constant, record the total counts for a radioactive source a period of 1 minute.	
C.2	2	Subtract background radiation from each to obtain the radiation from the source alone. For 2 and 5 minutes, subtract a background that is two times ($2\times$) and five times ($5\times$) the background count.	
C.2	3	Repeat steps 1 and 2 for 2 minutes and 5 minutes.	

Section	Step	Procedure	Check when finished
C.3	1	Use the results from C.2 to calculate the radiation received for 20 and 60 minutes.	

Section	Step	Procedure	Check when finished
C.4	1	By doubling your distance from a radioactive source, you receive one-fourth (1/4) the intensity of the radiation. See Figure 4.2. Place a radioactive source 1 m (100 cm) from the tube.	
C.4	2	Record the counts for 1 minute at 100 cm.	

Section	Step	Procedure	Check when finished
C.4	3	Repeat steps 1 and 3 for 75 cm, 50 cm, 25 cm, and 10 cm. Stop measuring the radiation if the count becomes too great for the operating level of the counter.	
C.4	4	Subtract the background radiation from each distance.	

Figure 4.2 The effect of radiation lessens as the distance from the source increases

Section	Step	Procedure	Check when finished
C.5	1	Calculate the ratio of the counts per minute at 50 and 100 cm. This will give the increase in radiation when the distance to the source is halved.	

Section	Step	Procedure	Check when finished
C.6	1	Graph the radiation level (cpm) against the distance of the source from the detection tube. Review instructions for preparing a graph found in the preface.	

Section	Step	Procedure	Check when finished
C.7	1	Using the graph, predict the counts per minute that would be obtained at distances of 20 and 40 cm.	

Laboratory 4

Nuclear Radiation

Report Sheet - Lab 4

Date _____ Name _____
Section _____ Team _____
Instructor _____   _____

Pre-Lab Study Questions

1. In what part of the atom do alpha or beta particles originate?

2. Why is protection from radiation needed?

3. What are some medical uses of radiation?

Laboratory 4

Nuclear Radiation

Report Sheet - Lab 4

Date _____ Name _____
Section _____ Team _____
Instructor _____ _____

A. Background Count

A.1 Counts during   minute 1   _____
                    minute 2   _____
                    minute 3   _____

A.2 Total counts   _____

    Average background count _____ counts/minute (cpm)

B. Radiation from Radioactive Sources

Item Tested	Counts per Minute (cpm)	Average Background (cpm)	Radiation from Source (cpm)	Type of Radiation

Questions and Problems

Q.1 Which item was the most radioactive?

Q.2 If that item is a consumer product, what is the source and purpose of the radioactive material?

Laboratory 4

Report Sheet - Lab 4

C. Effect of Shielding, Time, and Distance

C.1 Shielding

Shielding	Counts per Minute	Average Background (cpm)	Radiation from Source (cpm)
No shielding			

Questions and Problems

Q.3   Which type of shielding provides the best protection from radiation? Why?

C.2 and C.3   Time

Minutes	Counts	Average Background (cpm)	Radiation from Source
1			
2			
5			
C.3 20 (calculated)			
60 (calculated)			

Nuclear Radiation

Report Sheet - Lab 4

C.4   **Distance**

Distance	Counts per Minute	Average Background (cpm)	Radiation from Source (cpm)
100 cm			
75 cm			
50 cm			
25 cm			
10 cm			

C.5   Ratio = $\dfrac{\text{Counts/min at 50 cm}}{\text{Counts/min at 100 cm}}$ = _____

Questions and Problems

Q.4   Complete the nuclear equations by filling in the correct symbols:

$^{27}_{13}Al +$ ☐ $\rightarrow\ ^{24}_{11}Na +\ ^{4}_{2}He$

$^{131}_{53}I \rightarrow\ ^{0}_{-1}e +$ ☐

$^{96}_{40}Zr +$ ☐ $\rightarrow\ ^{1}_{0}n +\ ^{99}_{42}Mo$

Q.5   Write the symbols for the following types of radiation:

a. alpha particle _____   b. beta particle _____   c. gamma ray _____

Laboratory 4

Report Sheet - Lab 4

C.6　　Graph

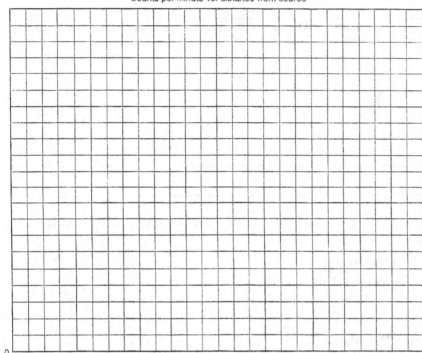

C.7　　Use your graph to estimate radiation levels at the following distances from the source:

　　　　Estimated cpm at 40 cm = _____

　　　　Estimated cpm at 20 cm = _____

Questions and Problems

Q.6　　List some shielding materials adequate for protection from:

　　　　a. alpha particles _____

　　　　b. beta particles _____

　　　　c. gamma rays _____

Q.7　　The bacteria in some foods are sterilized by placing the food near a source of ionizing radiation. Does that mean that the food becomes radioactive? Explain.

5
Chemical Reactions and Equations

Goals
- Observe physical and chemical properties associated with chemical changes.
- Give evidence for the occurrence of a chemical reaction.
- Write a balanced equation for a chemical reaction.
- Identify a reaction as a combination, decomposition, replacement, or combustion reaction.

Text References

Review the following before performing this laboratory.

Timberlake; *General, Organic, and Biological Chemistry: Structures of Life*, Third Edition
Sections 6.1 to 6.3; Pages 211 to 225

or

Timberlake; *Chemistry: An Introduction to General, Organic, and Biological Chemistry*, Tenth Edition
Section 5.3; Pages 171 to 184

Discussion

When a substance undergoes a physical change, it changes its appearance but not its composition. For example, when silver (Ag) melts and forms liquid silver (Ag), it undergoes a physical change from solid to liquid. In a chemical change, a substance is changed to give a new substance with a different composition and different properties. For example, when silver tarnishes, the shiny silver (Ag) changes to a dull-gray silver sulfide (Ag_2S), a new substance with different properties and a different composition. See Table 5.1.

Table 5.1 *Comparison of Physical and Chemical Changes*

Some Physical Changes	Some Chemical Changes
Change in state	Formation of a gas (bubbles)
Change in size	Formation of a solid (precipitates)
Tearing	Disappearance of a solid (dissolves)
Breaking	Change in color
Grinding	Heat is given off or absorbed

Laboratory 5

Balancing a Chemical Equation

In a chemical reaction, atoms in the reactants are rearranged to produce new combinations of atoms in the products. However, the total number of atoms of each element in the reactants is equal to the total number of atoms in the products. In an equation, the reactants are shown on the left and the products on the right. An arrow between them indicates that a chemical reaction takes place.

$$\text{Reactants} \rightarrow \text{Products}$$

To balance the number of atoms of each element on the left and right sides of the arrow, we write a number called a *coefficient* in front of the formula containing that particular element. Consider the balancing of the following unbalanced equation. The state of the substances as gas is shown using the parenthetical designation such as (g).

$N_2(g) + H_2(g) \rightarrow NH_3(g)$	*Unbalanced equation*
$N_2(g) + H_2(g) \rightarrow 2NH_3(g)$	*A coefficient of 2 balances the N atoms.*
$N_2(g) + 3H_2(g) \rightarrow 2NH_3(g)$	*A coefficient of 3 balances the H atoms.*
	The equation is now balanced.

Types of Reactions

There are many different chemical reactions, but most can be classified into the types of reactions shown in Table 5.2.

Table 5.2 *Common Types of Chemical Reactions*

Type of Reaction	Description	Example Equation
Combination	Elements or simple compounds form a more complex product.	$Cu + S \rightarrow CuS$
Decomposition	A reacting substance is split into simpler	$CaCO_3 \rightarrow CaO + CO_2$
Single replacement	One element takes the place of another element in a compound.	$Mg + 2HCl \rightarrow MgCl_2 + H_2$
Double replacement	Elements in two compounds switch places.	$AgNO_3 + NaCl \rightarrow AgCl + NaNO_3$
Combustion	Reactant and oxygen form an oxide product.	$S + O_2 \rightarrow SO_2$

Chemical Reactions and Equations

Lab Information

Time: 2–2 1/2 hr

Comments: Read all the directions and safety instructions carefully.

Match the labels on bottles and containers with the names of the substances you need.

Label your containers with the formulas of the chemicals you place in them.

Be sure that long hair is tied back.

A Bunsen burner is a potential hazard. Keep your work area clear of books, papers, backpacks, and other potentially flammable items.

Tear out the report sheets and place them beside the matching procedures.

Related Topics: Chemical change, chemical equation, balancing chemical equations

Experimental Procedures   *GOGGLES REQUIRED!*

A. Magnesium and Oxygen

Materials: Magnesium ribbon (2–3 cm long), tongs, Bunsen burner

Section	Step	Procedure	Check when finished
A.1	1	Obtain a small strip (2–3 cm) of magnesium ribbon.	
A.1	2	Record its appearance.	
A.1	3	Using a pair of tongs to hold the end of the magnesium ribbon, ignite it using the flame of a Bunsen burner. *As soon as the magnesium ribbon ignites, remove it from the flame. Shield your eyes as the ribbon burns and do not look directly at the flame.*	
A.1	4	Record your observations of the reaction and the physical properties of the product. Use complete sentences to describe your observations.	
A.2	1	Balance the equation given for the reaction. Use 1 as a coefficient when one unit of that substance is required. The letters in parentheses indicate the physical state of the reactant or product: (*g*) gas, (*s*) solid. *Unbalanced equation:* $Mg(s) + O_2(g) \rightarrow MgO(s)$	
A.3	1	Identify the type of reaction that has occurred. For this reaction, more than one reaction type may be used to classify the reaction.	

Laboratory 5

B. Zinc and Copper(II) Sulfate

Materials: Two test tubes, test tube rack, 1 M CuSO$_4$ (copper(II) sulfate solution), Zn(s)

For all the experiments in parts C–F, use small quantities. For solids, use the amount of compound that will fit on the tip of a spatula or small scoop. Carefully pour small amounts of liquids into your own beakers and other containers. Measure out 3 mL of water in a test tube. Use this volume as a reference level for each of the experiments.

Do not place droppers or stirring rods into reagent bottles. They may contaminate a reagent for the entire class. Discard unused chemicals as indicated by your instructor.

Section	Step	Procedure	Check when finished
B.1	1	Pour 3 mL (match the reference volume) of the *blue solution*, 1 M CuSO$_4$ (one molar copper(II) sulfate), into each of two test tubes.	
B.1	2	Obtain a small piece of zinc metal.	
B.1	3	Describe the appearance of the CuSO$_4$ solution and the small piece of zinc metal.	
B.1	4	Add the Zn metal piece to the CuSO$_4$ solution in one of the test tubes. The CuSO$_4$ solution in the other test tube is your reference for the initial solution color.	
B.1	5	Place the test tubes in your test tube rack and observe the color of the CuSO$_4$ solutions and the Zn piece again at 15 and 30 minutes.	
B.1	6	Pour the CuSO$_4$ solutions into a waste contains or dispose of as directed by your instructor.	
B.1	7	Rinse the piece of zinc with water and place it in a recycling container or as directed by your instructor.	

Section	Step	Procedure	Check when finished
B.2	1	Balance the equation given for the reaction. The symbol (*aq*) means aqueous (dissolved in water). Unbalanced equation: Zn(s) + CuSO$_4$(aq) → Cu(s) + ZnSO$_4$(aq)	
B.3	1	Identify the type of reaction that has occurred.	

Chemical Reactions and Equations

C. Metals and HCl

Materials: Three test tubes, test tube rack, small pieces of Cu(s), Zn(s), and Mg(s) metal

1 M HCl *Caution: HCl is a corrosive acid. Handle carefully!*

Section	Step	Procedure	Check when finished
C.1	1	Place 3 mL of 1 M HCl (match your reference volume from part C) in each of three test tubes.	
C.1	2	Describe the appearance of each metal.	
C.1	3	Carefully add a metal piece to the acid in each of the test tubes.	
C.1	4	Record any evidence of reaction such as bubbles of gas (H_2).	
C.1	5	Carefully pour off the acid and follow with large quantities of water to dilute. Dispose of as directed by your instructor.	
C.1	6	Rinse the metal pieces with water, dry, and return to your instructor.	

Section	Step	Procedure	Check when finished
C.2	1	Balance the equation given for each metal that gave a chemical reaction. If there was no reaction, cross out the products and write NR for no reaction. *Unbalanced equations:* 1. $Cu(s) + HCl(aq) \rightarrow CuCl_2(aq) + H_2(g)$ 2. $Zn(s) + HCl(aq) \rightarrow ZnCl_2(aq) + H_2(g)$ 3. $Mg(s) + HCl(aq) \rightarrow MgCl_2(aq) + H_2(g)$	
C.3	1	Identify the type of reaction for each chemical reaction that occurred.	

D. Reactions of Ionic Compounds

Materials: Three (3) test tubes, test tube rack, dropper bottle sets of 0.1 M solutions: $CaCl_2$, Na_3PO_4, $BaCl_2$, Na_2SO_4, $FeCl_3$, KSCN

For each of these reactions, two substances will be mixed together. Describe your observations of the reactants before you mix them and then describe the products of the reaction. Look for changes in color, the formation of a solid (solution turns cloudy), the dissolving of a solid, and/or the formation of a gas (bubbling). Balance the equations for the reactions. Dispose of the solutions properly as directed by your instructor.

Laboratory 5

Section	Step	Procedure	Check when finished
D.1	1	Place 20 drops each of 0.1 M $CaCl_2$ (calcium chloride) and 0.1 M Na_3PO_4 (sodium phosphate) into a test tube.	
D.1	2	Describe any changes that occur.	
D.1	3	Identify the type of reaction for each chemical reaction that occurred. *Unbalanced equation:* $CaCl_2(aq) + Na_3PO_4(aq) \rightarrow Ca_3(PO_4)_2(s) + NaCl(aq)$	

Section	Step	Procedure	Check when finished
D.2	1	Place 20 drops each of 0.1 M $BaCl_2$ (barium chloride) and 0.1 M Na_2SO_4 (sodium sulfate) into a test tube.	
D.2	2	Describe any changes that occur.	
D.2	3	Identify the type of reaction for each chemical reaction that occurred. Unbalanced equation: $BaCl_2(aq) + Na_2SO_4(aq) \rightarrow BaSO_4(s) + NaCl(aq)$	

Section	Step	Procedure	Check when finished
D.3	1	Place 20 drops each of 0.1 M $FeCl_3$ (iron(III) chloride) and 0.1 M KSCN (potassium thiocyanate) into a test tube.	
D.3	2	Describe any changes that occur.	
D.3	3	Identify the type of reaction for each chemical reaction that occurred. *Unbalanced equation:* $FeCl_3(aq) + KSCN(aq) \rightarrow Fe(SCN)_3(aq) + KCl(aq)$	

E. Sodium Carbonate and HCl

Materials: Test tube, test tube rack, 1 M HCl solution, $Na_2CO_3(s)$, and matches or wood splints

Section	Step	Procedure	Check when finished
E.1	1	Place about 3 mL of 1 M HCl in a test tube.	
E.1	2	Add a small amount of solid Na_2CO_3 (about the size of a pea) to the test tube.	
E.1	3	Identify the type of reaction for each chemical reaction that occurred. Unbalanced equation: $FeCl_3(aq) + KSCN(aq) \rightarrow Fe(SCN)_3(aq) + KCl(aq)$	
E.1	4	Record your observations. *Caution: HCl is corrosive. Clean up any spills immediately. If spilled on the skin, flood the area with water for at least 10 minutes.*	
E.2	1	Identify the type of reaction for each chemical reaction that occurred. *Unbalanced equation:* $Na_2CO_3(s) + HCl(aq) \rightarrow CO_2(g) + H_2O(l) + NaCl(aq)$	

Section	Step	Procedure	Check when finished
E.3	1	Light a match or wood splint and insert the flame inside the neck of the test tube. What happens to the flame?	
E.3	2	Record your observations.	

Laboratory 5

Chemical Reactions and Equations

Report Sheet - Lab 5

Date _____   Name _____
Section _____   Team _____
Instructor _____   _____

Pre-Lab Study Questions

1. Why is the freezing of water called a physical change?

2. Why are burning candles and rusting nails examples of chemical change?

3. What is included in a chemical equation?

4. How does a combination reaction differ from a decomposition reaction?

Laboratory 5

Chemical Reactions and Equations

Report Sheet - Lab 5

Date _____ Name _____
Section _____ Team _____
Instructor _____   _____

A. Magnesium and Oxygen

A.1   Initial appearance of Mg   _____

　　　Observations of the reaction   _____

　　　Appearance of the product   _____

A.2   Balance: ____Mg(s) + ____O$_2$(g) → ____MgO(s)

A.3   Type of reaction: _____

B. Zinc and Copper(II) Sulfate

B.1   Initially Zn   _____

　　　　　　　　 CuSO$_4$ _____

　　　15 min Zn   _____

　　　　　　　　 CuSO$_4$ _____

　　　30 min Zn   _____

　　　　　　　　 CuSO$_4$ _____

B.2   Balance: ____Zn(s) + ____CuSO$_4$(aq) → ____Cu(s) + ____ZnSO$_4$(aq)

B.3   Type of reaction: _____

C. Metals and HCl

C.1   Observations

Cu   Initial: _____

　　　Reaction: _____

63

Laboratory 5

Report Sheet - Lab 5

Zn   Initial: _____

         Reaction: _____

Mg   Initial: _____

         Reaction: _____

C.2   Balance: ____Cu(s) + ____HCl(aq) → ____CuCl$_2$(aq) + ____H$_2$(g)

                    ____Zn(s) + ____HCl(aq) → ____ZnCl$_2$(aq) + ____H$_2$(g)

                    ____Mg(s) + ____HCl(aq) → ____MgCl$_2$(aq) + ____H$_2$(g)

C.3   Type of reaction: Cu _____

                          Zn _____

                          Mg _____

D.   Reactions of Ionic Compounds

D.1   **CaCl$_2$ and Na$_3$PO$_4$**

         Observations: _____

         Type of reaction: _____

         Balance: ____CaCl$_2$(aq) + ____Na$_3$PO$_4$(aq) → ____Ca$_3$(PO$_4$)$_2$(s) + ____NaCl(aq)

D.2   **BaCl$_2$ and Na$_2$SO$_4$**

         Observations: _____

         Type of reaction: _____

         Balance: ____BaCl$_2$(aq) + ____Na$_2$SO$_4$(aq) → ____BaSO$_4$(s) + ____NaCl(aq)

Chemical Reactions and Equations

Report Sheet - Lab 5

D.3 **FeCl₃ and KSCN**

Observations: _____

Type of reaction: _____

Balance: ____ $FeCl_3(aq)$ + ____ $KSCN(aq)$ → ____ $Fe(SCN)_3(aq)$ + ____ $KCl(aq)$

E. Sodium Carbonate and HCl

E.1 Observations: _____

E.2 Type of reaction: _____

Balance: ____ $Na_2CO_3(s)$ + ____ $HCl(aq)$ → ____ $CO_2(g)$ + ____ $H_2O(l)$ + ____ $NaCl(aq)$

E.3 Why did the flame of the burning match or splint go out?

Questions and Problems

Q.1 What evidence of a chemical reaction might you see in the following cases?

a. Dropping an Alka-Seltzer™ tablet into a glass of water

_____

b. Bleaching a stain

_____

c. Burning a match

_____

d. Rusting of an iron nail

_____

Laboratory 5

Report Sheet - Lab 5

Q.2 Balance the following equations:

a. ___ Mg(s) + ___ HCl(aq) → ___ H_2(g) + ___ $MgCl_2$(aq)

b. ___ Al(s) + ___ O_2(g) → ___ Al_2O_3(s)

c. ___ Fe_2O_3(s) + ___ H_2O(l) → ___ $Fe(OH)_3$(s)

d. ___ $Ca(OH)_2$(aq) + ___ HNO_3(aq) → ___ $Ca(NO_3)_2$(aq) + ___ H_2O(l)

Q.3 Write an equation for the following reactions. Remember that gases of elements such as oxygen are diatomic (O_2). Write the *correct formulas* of the reactants and products. Then correctly balance each equation.

a. Potassium and oxygen gas react to form potassium oxide.

b. Sodium and water react to form sodium hydroxide and hydrogen gas.

c. Iron and oxygen gas react to form iron(III) oxide.

Q.4 Classify each reaction as combination (C), decomposition (DC), single replacement (SR), or double replacement (DR).

a. Ni + F_2 → NiF_2 _____

b. Fe_2O_3 + 3C → 2Fe + 3CO _____

c. $CaCO_3$ → CaO + CO_2 _____

d. H_2SO_4 + 2KOH → K_2SO_4 + $2H_2O$ _____

Q.5 Predict what product(s) would form from the reaction of the following reactants:

a. Zn + $CuBr_2$ → _____ + _____

b. H_2 + Cl_2 → _____

c. $MgCO_3$ → _____ + _____

d. KCl + $AgNO_3$ → _____ + _____

6

Moles and Chemical Formulas

Goals
- Use the mole conversion factors to convert grams to moles and moles to grams.
- Experimentally determine the simplest formula of an oxide of magnesium.
- Calculate the percent water in a hydrated salt.
- Determine the formula of a hydrate.

Text References

Review the following before performing this laboratory.

Timberlake; *General, Organic, and Biological Chemistry: Structures of Life*, Third Edition
Sections 6.4 to 6.8; Pages 226 to 246

or

Timberlake; *Chemistry: An Introduction to General, Organic, and Biological Chemistry*, Tenth Edition
Sections 5.1 to 5.2; Pages 162 to 171

Discussion

A. Finding the Simplest or Empirical Formula

In this experiment, magnesium metal is heated to a high temperature until it reacts with the oxygen (O_2) in the air.

Reactants	Heat	**Product**
$2Mg(s) + O_2(g)$	$\rightarrow$	$2MgO(s)$
silvery metal		*white-gray ash*

As the magnesium burns in the air, it may also combine with the nitrogen (N_2) in the air. To remove any nitride product, water is added and the product is reheated. Any nitride product is converted to magnesium oxide and ammonia.

$$3Mg(s) + N_2(g) \xrightarrow{\text{Heat}} Mg_3N_2(s)$$

$$Mg_3N_2(s) + 3H_2O(l) \xrightarrow{\text{Heat}} 3MgO(s) + 2NH_3(g)$$

Laboratory 6

At the beginning of the experiment, you will obtain the mass of the magnesium metal. At the end, you will determine the mass of the oxide product. The mass of oxygen that combined with the magnesium is the difference between the mass of the oxide product and the original mass of magnesium.

g oxide product − g Mg = g O that combined to form the oxide product

The simplest formula of the magnesium oxide product is determined by calculating the moles of magnesium and the moles of oxygen using their respective molar masses. This is often called the empirical formula.

$$moles\ Mg = g\ Mg\ ribbon \times \frac{1\ mole}{24.3\ g\ Mg}$$

$$moles\ O = grams\ of\ O\ combined \times \frac{1\ mole\ O}{16.0\ g\ O}$$

The simplest or empirical formula is obtained by dividing the number of moles of the elements by the smaller number of moles. Suppose that in a similar experiment we determined that 0.040 mole of Zn had combined with 0.080 mole of Cl to form a compound. We can proceed as follows:

1. Ratio of moles: 0.040 mole Zn : 0.080 mole Cl.

2. Determine the smaller number of moles: 0.040 mole Zn.

3. Divide the moles of each element by the smaller number of moles and round to the nearest whole number.

$$\frac{0.080\ mole\ Cl}{0.040} = 2\ moles\ Cl \qquad \frac{0.040\ mole\ Zn}{0.040} = 1\ mole\ Zn$$

4. Use the whole numbers as subscripts to write the formula of the compound.

   Zn_1Cl_2 or $ZnCl_2$

B. Formula of a Hydrate

A *hydrate* is a salt that contains a specific number of water molecules called the *water of hydration*. The number of water molecules is fixed for each hydrate, but differs from one salt to another. In the formula of a hydrate, the number of water molecules is written after the salt formula and is separated by a large, raised dot.

$CaSO_4 \cdot H_2O$ $CuSO_4 \cdot 5H_2O$ $Na_2CO_3 \cdot 10H_2O$

Heating the hydrate provides the energy to remove the water molecules. The salt without water is called an anhydrate.

$$CuSO_4 \cdot 5H_2O \quad \xrightarrow{Heat} \quad CuSO_4 + 5\ H_2O(g)$$

hydrate   anhydrate   water of hydration

Moles and Chemical Formulas

Lab Information

Time: 2–2½ hr

Comments: Tear out the report sheets and place them beside the matching procedures.

Use steel wool to remove any dull coating on the magnesium ribbon until it is shiny.

Check the crucible for cracks before you start to heat it.

When you set a hot object aside to cool, remember that it is hot.

Dispose of all chemicals as directed by your instructor.

Related Topics: Formulas, moles, molar mass, calculating moles from grams, calculating grams from moles, hydrates, dehydration

Experimental Procedures *GOGGLES REQUIRED!*

A. Finding the Simplest Formula

Materials: Crucible, crucible cover, crucible tongs, clay triangle, iron ring, ring stand, Bunsen burner, magnesium ribbon, steel wool, eyedropper, small 100- or 150-mL beaker, laboratory balance, heat-resistant pad

Section	Step	Procedure	Check when finished
A.1	1	Obtain a clean, dry crucible and its cover. (A porcelain crucible may have stains in the porcelain that cannot be removed.)	
A.1	2	Place the crucible and cover in a clay triangle that is sitting on an iron ring attached to a ring stand. See Figure 6.1.	
A.1	3	Heat for 1 minute.	
A.1	4	Cool (5–10 minutes) until the crucible is at room temperature.	
A.1	5	Using crucible tongs, carry the crucible and cover to the balance.	
A.1	6	Weigh the crucible and cover (together) and record. Give all the figures in the mass; do not round off. ***Do not place hot objects on a balance pan.***	

Laboratory 6

Figure 6.1 Heating a crucible and cover in a clay triangle

Section	Step	Procedure	Check when finished
A.2	1	Obtain a piece of magnesium ribbon that has a mass of 0.15–0.3 g.	
A.2	2	If there is tarnish on the ribbon, remove it by polishing the ribbon with steel wool.	
A.2	3	Describe the appearance of the magnesium ribbon.	
A.2	4	Wind the ribbon into a coil and place the coil into a crucible.	
A.2	5	Weigh the crucible and cover with the magnesium.	
A.2	6	Record the total mass. *Do not round off mass.*	
A.2	7	**Do this part of the experiment in a fume hood:** Place the crucible and magnesium ribbon in a clay triangle. Have the cover and a pair of tongs on the desktop.	
A.2	8	Begin to heat the crucible.	
A.2	9	Watch closely for smoke or fumes, which is the first indication of the magnesium and oxygen reacting. If the magnesium is slow to ignite, fan some air into the crucible.	
A.2	10	When the magnesium bursts into flame, use tongs to replace the cover on the crucible.	
A.2	11	If the magnesium does not burn, check the flame. The tip of the inner flame must touch the bottom of the crucible. The bottom of the crucible will be red hot. *Caution: As soon as the magnesium begins to smoke or bursts into flame, place the cover over the crucible using your tongs. Avoid looking directly at the bright flame of the burning magnesium.*	
A.2	12	Observe the reaction of the magnesium and oxygen about every minute.	

Moles and Chemical Formulas

Section	Step	Procedure	Check when finished
A.2	13	When the magnesium *no longer* produces smoke or a flame, remove the cover and place it on a heat-resistant surface.	
A.2	14	Heat the crucible strongly for another 5 minutes.	
A.2	15	Turn off the burner and let the crucible and its contents cool to room temperature. The crucible can be moved to the heat-resistant surface using crucible tongs. ***Caution: The crucible and iron ring are still hot.***	
A.2	16	During the heating, some of the magnesium reacts with nitrogen in the air to form a magnesium nitride product. To remove this nitride product, *carefully* add 15–20 drops of water to the *cooled* contents.	
A.2	17	Heat the covered crucible and its contents gently for 5 minutes to evaporate the water. Do not allow the water to splatter as then you may lose some of the product.	
A.2	18	Then heat strongly for 5 minutes. ***Caution: Avoid breathing fumes from the crucible because ammonia may be released during the heating.***	

Section	Step	Procedure	Check when finished
A.3	1	Allow the crucible to cool. You may wish to move the crucible and cover to a heat-resistant surface where it can cool completely.	
A.3	2	Reweigh the crucible, cover, and oxide contents.	
A.3	3	Record their total mass.	
A.3	4	Describe the appearance of the oxide product.	

Calculations

Section	Step	Procedure	Check when finished
A.4	1	Determine the mass of the magnesium ribbon present originally. (A.2 – A.1)	
A.5	1	Calculate the mass of the magnesium oxide product. (A.3 – A.1)	
A.6	1	Calculate the mass of oxygen that combined with the magnesium by subtracting the mass of Mg originally present from the final mass of the MgO product. g MgO product (A.5) – g Mg (A.4) = g O (reactant)	

Laboratory 6

A.7	1	Determine the number of moles of magnesium present in the ash product by dividing the mass of the magnesium by its molar mass. g Mg (A.4) x $\frac{1 \text{ mole Mg}}{24.3 \text{ g Mg}}$	
A.8	1	Determine the number of moles of oxygen present in the ash product by dividing the mass of the oxygen by its molar mass. g O (A.6) x $\frac{1 \text{ mole O}}{16 \text{ g O}}$	
A.9	1	In their chemical formula, the moles of magnesium in proportion to the moles of oxygen are stated as small whole numbers. Now we need to change the calculated moles of magnesium and oxygen to whole numbers. To do this, divide both values by the smaller one. That makes the smaller value 1. If our lab work and weighing have been carefully done, the other number should be very close to another whole number such as 1, 2	
A.9	2	Round the results to the nearest whole number.	
A.10	1	Use the whole number values obtained in A.9 as subscripts in writing the simplest formula of the product of magnesium and oxygen.	

B. Formula of a Hydrate

Materials: Crucible, clay triangle, crucible tongs, hydrate of $MgSO_4$, iron ring and stand, Bunsen burner, heat-resistant pad, laboratory balance

Section	Step	Procedure	Check when finished
B.1	1	Obtain a clean, dry crucible.	
B.1	2	Heat it for 2–3 minutes.	
B.1	3	Let it cool.	
B.1	4	Weigh carefully.	
B.1	5	Record the weight.	

Section	Step	Procedure	Check when finished
B.2	1	Fill the crucible about 1/3 full with a hydrate of $MgSO_4$.	
B.2	2	Weigh the crucible and the $MgSO_4$.	
B.2	3	Record the total mass of the crucible and the hydrate.	

Moles and Chemical Formulas

Section	Step	Procedure	Check when finished
B.3	1	Set the crucible and hydrate on a clay triangle that is set on an iron ring. See Figure 9.1.	
B.3	2	Heat gently for 5 minutes.	
B.3	3	Then increase the intensity of the flame and heat strongly for 10 minutes more. The bottom of the crucible should become a dull red color.	
B.3	4	Turn off the burner.	
B.3	5	Using crucible tongs, move the crucible to a heat-resistant surface.	
B.3	6	After the crucible cools to room temperature, weigh the crucible and its contents. *Caution: Allow heated items to cool to room temperature. Do not place a hot container on a balance pan.*	
B.3	7	Record the mass if it is the final mass. (See optional steps below.)	
B.3	8	**Optional** To be sure that you have completely driven off the water of hydration, heat the crucible and its contents for another 5 minutes.	
B.3	9	Cool.	
B.3	10	Reweigh. If the mass in the second heating is within 0.05 g of the mass obtained after the first heating, you have completely dehydrated the salt. If not, heat again until you have agreement between final masses.	
B.3	11	Record the final mass and use for calculation.	

Calculations

Section	Step	Procedure	Check when finished
B.4	1	Calculate the mass of the hydrate present.	
B.5	1	Calculate the mass of the anhydrous product after heating.	
B.6	1	Calculate the mass of water released from the hydrate sample by subtracting the mass of the anhydrous product from the original mass of the hydrate. g salt hydrate (B.4) − g anhydrate (B.5) = g H_2O	
B.7	1	Calculate the percent H_2O in the hydrate by dividing the mass of H_2O by the original mass of the hydrate. $$\frac{g\ water\ (B.6)}{g\ hydrate\ (B.4)} \times 100 = \%water\ in\ hydrate$$	

Laboratory 6

B.8	1	Calculate the moles of H$_2$O in the hydrate by dividing the mass of the H$_2$O by its molar mass. $$\frac{1 \text{ mole}}{18.0 \text{ g water}} \times g \text{ water } (B.6) = \text{moles water}$$	
B.9	1	Calculate the moles of anhydrate by dividing the mass of the anhydrate by its molar mass. $$\frac{1 \text{ mole anhydrate}}{120.4 \text{ g anhydrate}} \times g \text{ anhydrate } (B.5) = \text{moles anhydrate}$$	
B.10	1	To determine the formula of the hydrate we need the following ratio: _____ moles of water to 1 mole of anhydrate To determine the ratio of moles of water to 1 mole of anhydrate, divide the moles of water (see B.8) by the moles of anhydrate (B.9). Round off the value for moles of H$_2$O to the nearest whole number. $$\frac{\text{Moles water } (B.8)}{\text{Moles anhydrate } (B.9)} = \frac{\text{Moles water}}{1 \text{ mole anhydrate}}$$	
B.11	1	Complete the formula of your hydrate by writing in the number of moles of water.	

Moles and Chemical Formulas

Report Sheet - Lab 6

Date _____ Name _____
Section _____ Team _____
Instructor _____   _____

Pre-Lab Study Questions

1. What is meant by the simplest formula of a compound?

2. How does a hydrate differ from an anhydrate?

3. What happens when a hydrate is heated?

Laboratory 6

Report Sheet - Lab 6

Date _____ Name _____
Section _____ Team _____
Instructor _____ _____

A. Finding the Simplest Formula

A.1 Mass of empty crucible + cover _____ g

A.2 Initial appearance of magnesium _____

     Mass of crucible + cover + magnesium _____ g

A.3 Mass of crucible + cover + oxide product _____ g

     Appearance of oxide product _____

Calculations

A.4 Mass of magnesium _____ g Mg

A.5 Mass of magnesium oxide product _____ g product

A.6 Mass of oxygen in the product _____ g O

A.7 Moles of Mg _____ mole Mg

(Show calculations.)

Laboratory 6

Report Sheet - Lab 6

A.8 Moles of O _____ mole O

(Show calculations.)

A.9 _____ mole Mg : _____ mole O : smaller value _____

  moles Mg = _____ moles Mg (rounded)

  moles O = _____ moles O (rounded)

A.10 Formula: Mg☐ O☐ ← subscripts

Questions and Problems

Q.1 Using the rules for writing the formulas of ionic compounds, write the ions and the correct formula for magnesium oxide.

Q.2 Write a balanced equation for the reaction of the magnesium and the oxygen including their physical states.

Moles and Chemical Formulas

Report Sheet - Lab 6

Q.3   How many grams are needed to obtain 1.5 moles of magnesium oxide?

Q.4   How many moles are in each of the following?

a. 80.0 g of NaOH

b. 12.0 g of Ca(OH)$_2$

Q.5   Write the simplest formula for each of the following compounds:

0.200 mole Al and 0.600 mole Cl

0.080 mole Ba, 0.080 mole S, 0.320 mole O

Laboratory 6

Report Sheet - Lab 6

Q.6 When 2.50 g of copper (Cu) reacts with oxygen, the copper oxide product has a mass of 2.81 g. What is the simplest formula of the copper oxide?

Q.7 Aluminum and oxygen gas react to produce aluminum oxide.

a. Write a balanced equation for the reaction.

b. If 12 moles of Al_2O_3 are produced, how many moles of aluminum reacted?

c. If 75 g of oxygen react, how many grams of aluminum are required?

B.   Formula of a Hydrate

B.1 Mass of crucible _____ g

B.2 Mass of crucible and salt (hydrate) _____ g

B.3 Mass of crucible and salt (anhydrate) after first heating _____ g

after second heating (optional) _____ g

Moles and Chemical Formulas

Report Sheet - Lab 6

Calculations

B.4   Mass of salt (hydrate)   _____ g

B.5   Mass of salt (anhydrate)   _____ g

B.6   Mass of water lost   _____ g

B.7   Percent water   _____ %

(Show calculations.)

B.8   Moles of water   _____ mole

(Show calculations.)

B.9   Moles of salt (anhydrate)   _____ moles

(Show calculations.)

B.10   Ratio of moles of water to moles of hydrate

$$\frac{_____ \text{ moles water (B.8)}}{_____ \text{ moles MgSO}_4 \text{ (B.9)}} = \frac{_____ \text{ moles H}_2\text{O}}{1 \text{ mole MgSO}_4}$$

Laboratory 6

Report Sheet - Lab 6

B.11   Formula of hydrate (salt) $MgSO_4 \cdot \square\ H_2O$

Questions and Problems

Q.8   Using the formula you obtained in B.11, write a balanced equation for the dehydration of the $MgSO_4$ hydrate you used in the experiment.

7

Energy and Matter

Goals

- Prepare a heating curve and a cooling curve.
- Use the specific heat of water to calculate heat lost or gained.
- Calculate the heat of fusion for water.
- Calculate the caloric values of foods in kcal/g to calculate the kilocalories in a serving of food.
- Use nutrition data on food products to determine the kilocalories in one serving.
- Identify a reaction as exothermic or endothermic.

Text Reference

Review the following before performing this laboratory.

Timberlake; *General, Organic, and Biological Chemistry: Structures of Life*, Third Edition
Sections 2.3 to 2.7 and 6.9 to ; Pages 63 to 86 and 246 to 248

or

Timberlake; *Chemistry: An Introduction to General, Organic, and Biological Chemistry*, Tenth Edition
Sections 2.1, 2.2, and 2.4 to 2.6 and 5.9; Pages 52 to 58 and 63 to 73 and 195 to 197

Discussion

A. A Heating Curve for Water

The temperature of a substance indicates the kinetic energy (energy of motion) of its molecules. When water molecules gain heat energy, they move faster and the temperature rises. Eventually the water molecules gain sufficient energy to separate from the other liquid molecules. The liquid changes to a gas in a change of state called *boiling*. The change of state from liquid to gas is indicated when the water temperature becomes constant. It is more obvious when a graph is drawn of the temperature change of the substance that is heated. When a liquid boils, a horizontal line (*plateau*) appears on the graph, as shown in Figure 7.1. This constant temperature is called its *boiling point*.

Laboratory 7

Figure 7.1 Example of a graph of a heating curve

Measuring Heat

When two substances are in contact, the heat from one can be transferred to the other. When you put ice in a warm drink, the heat from the drink is transferred to the ice. The ice melts and the drink cools. The drink lost heat and the ice gained heat. By measuring the change in temperature of the drink, we can determine the amount of heat transferred to the ice.

$$\text{Heat lost (by warm drink)} = \text{heat gained (by ice)}$$

The amount of heat transferred depends on the specific heat of the substance. Specific heat refers to the amount of heat required to raise the temperature of 1 g of a substance by 1°C.

$$\text{Specific heat of water} = \frac{1.00 \text{ calorie}}{1 \text{ g} \times 1°C}$$

Water has a specific heat of 1.00 calorie per g °C. The heat energy (calories) is calculated using the mass (g) of water, the change in temperature (ΔT), and the specific hear of water, according to the following:

$$\text{Heat (cal)} = \text{mass of water (g)} \times \text{temperature change} (\Delta T) \times \text{specific heat} (1.00 \text{ cal/ g °C})$$

B. Energy in Changes of State

Heat of Fusion

Changing state from solid to liquid (melting) requires energy. Ice melts at a constant temperature of 0°C. At that melting point, the amount of heat required to melt 1 g of ice is called the *heat of fusion*. For water, the energy needed to melt 1 g of ice (at 0°C) is 80. calories. That is also the amount of heat released when liquid water freezes to solid ice at 0°C.

Melting (0°C): $H_2O(s) + \text{heat}_{\text{fusion}}$ (80. cal/g) $\rightarrow H_2O(l)$

Freezing (0°C): $H_2O(l) \rightarrow H_2O(s) + \text{heat}_{\text{fusion}}$ (80. cal/g)

In this experiment, ice will be added to a sample of water. From the temperature change, the amount of heat lost by the water sample can be calculated. This is also the amount of heat needed to melt the ice.

Heat (cal) lost by water = heat (cal) gained to melt ice = g water × ΔT × 1.00 cal/g °C

By measuring the amount of ice that melted, the heat of fusion can be calculated as follows:

$$\text{Heat of fusion (cal/g)} = \frac{\text{heat (cal) gained to melt ice}}{\text{grams of ice}}$$

Heat of Vaporization

A similar situation occurs for a substance that changes from a liquid to a gas (vapor). Water boils at 100°C, its boiling point. At that temperature, the energy required to convert liquid to gas is called the *heat of vaporization*. For water, the energy required to vaporize 1 g of water at 100°C is 540 calories. This is also the amount of heat released when 1 gram of steam condenses to liquid at 100°C.

Boiling (100°C): $H_2O(l)$ + heat$_{vaporization}$ (540 cal/g) → $H_2O(g)$

Condensation (100°C): $H_2O(g)$ → $H_2O(l)$ + heat$_{vaporization}$ (540 cal/g)

C. Food Calories

Our diets contain foods that provide us with energy. We need energy to make our muscles work, to breathe, to synthesize molecules in the body such as protein and fats, and to repair tissues. A typical diet required by a 25-year-old woman is about 2000–2500 kcal. By contrast a bicycle rider has a higher energy requirement and needs a diet that provides 4000 kcal. The nutritional energy of food is determined in Calories (Cal), which are the same as 1000 cal or 1 kilocalorie.

Nutritionists use calorimeters to establish the caloric values for the three food types: carbohydrates, 4 kcal/g; fats, 9 kcal/g; and proteins, 4 kcal/g. By measuring the amount of each food type in a serving, the kilocalories can be calculated. For example, a candy that is composed of 12 g of carbohydrate will provide 48 kcal. Usually the values are rounded to the nearest tens place.

$$12 \text{ g carbohydrate} \times \frac{4 \text{ kcal}}{1 \text{ g carbohydrate}} = 48 \text{ kcal or } 50 \text{ kcal}$$

D. Exothermic and Endothermic Reactions

In an *exothermic* reaction, heat is released, which causes the temperature of the surroundings to increase. An *endothermic* reaction absorbs heat, which causes a drop in the temperature of the surroundings. Heat can be written as a product in an equation for an exothermic reaction and as a reactant in the equation for an endothermic reaction. Energy is required to break apart bonds and is released when bonds form. If the energy released by forming bonds is greater than the energy required to break bonds apart, the reaction is exothermic. If the energy required to break apart bonds is greater than the energy released with forming bonds, the reaction is endothermic. In our cells, the bonds in carbohydrates are broken down to give us energy. Reactions that build molecules and repair cells are endothermic because they require energy.

Laboratory 7

Exothermic reactions: $C + O_2 \longrightarrow CO_2 + \text{heat}$

$C_6H_{12}O_6 + 6O_2 \longrightarrow 6CO_2 + 6H_2O + \text{energy}$

Glucose

Endothermic reactions: $\text{Heat} + PCl_5 \longrightarrow PCl_3 + Cl_2$

Energy + amino acids $\longrightarrow$ protein

Lab Information

Time:	2 hr
Comments:	Tear out the report sheets and place them beside the matching procedures.
	Be careful with boiling water.
	Use mitts or beaker tongs to move hot beakers, or let them cool.
Related Topics:	Changes of state, heating and cooling curves, heat of fusion, heat of vaporization

Experimental Procedures *GOGGLES REQUIRED!*

A. A Heating Curve for Water

Materials: Beaker (250- or 400-mL), Bunsen burner (or hot plate), ring stand, graduated cylinder, iron ring, wire gauze, clamp, thermometer, timer

Section	Step	Procedure	Check when finished
A	1	Using a graduated cylinder, pour 100 mL of cool water into a 250-mL beaker.	
A	2	As shown in Figure 7.2, place the beaker on a hot plate or on a wire screen placed on an iron ring above a Bunsen burner. The height of the iron ring should be about 3–5 cm above the burner.	
A	3	Tie a string to the loop in the top of the thermometer or place the thermometer securely in a clamp.	
A	4	Adjust the thermometer so that the bulb is in the center portion of the liquid. (Do not let the thermometer rest on the side or bottom of the beaker.)	

86

Energy and Matter

Figure 7.2 Setup for heating water with (left) a hot plate or (right) a Bunsen burner

Section	Step	Procedure	Check when finished
A.1	1	Measure and record the initial temperature of the water.	
A.1	2	Light the burner (or use a hot plate).	
A.1	3	Using a timer or a watch with a second hand, record the temperature of the water at 1-minute intervals. Eventually the water will come to a full boil. (The early appearance of small bubbles of escaping gas does not indicate boiling.)	
A.1	4	When the water is boiling, the temperature will become constant. Record the boiling temperature for another 4–5 minutes.	

Section	Step	Procedure	Check when finished
A.2	1	Prepare a heating curve for water by graphing the temperature versus the time. For a review of graphing, see page xviii in the preface of this lab manual.	
A.2	2	Label the parts of the graph that represent the liquid state and boiling.	
A.3	1	The plateau (flat part of graph) indicates the boiling point of the water. Record its value.	
A.4	1	Calculate the temperature change for the cool water to reach the boiling point (plateau).	
A.5	1	Using the measured volume of water and its density (1.00 g/mL), calculate the mass of the water. A (mL) × 1.00 g/mL	

Laboratory 7

A.6	1	Calculate the heat in calories used to heat the water to the boiling point. Calories = mass (A.5) × ΔT(A.4) × 1.00 cal/g °C	

B. Energy in Changes of State

Materials: Calorimeter (Styrofoam cup and cardboard cover), thermometer, 50- or 100-mL graduated cylinder, 100-mL beaker, ice

Section	Step	Procedure	Check when finished
B.1	1	Weigh an empty Styrofoam cup.	
B.2	1	Add 100 mL of water to the cup and reweigh.	

Section	Step	Procedure	Check when finished
B.3	1	Record the initial temperature of the water in the calorimeter. See Figure 7.3.	
B.3	2	Add 2 or 3 ice cubes (or crushed ice that fills a 100-mL beaker) to the water in the cup.	
B.3	3	Stir vigorously.	
B.3	4	Check the temperature of the ice water.	
B.3	5	Add ice until the temperature drops to 2–3°C.	
B.3	6	If some ice is not melted, remove it immediately.	
B.3	7	Record the final temperature of the water.	

Figure 7.3 Calorimetry setup with water, a thermometer, Styrofoam cup, and a cardboard cover

Calculations

Section	Step	Procedure	Check when finished
B.4	1	Weigh the Styrofoam (calorimetry) cup with the initial sample of water and the melted ice. The increase in mass indicates the amount of ice that melted.	
B.5	1	Calculate the mass of water added to the Styrofoam cup. Mass of water + cup (C.2) – Mass of cup (C.1)	
B.6	1	Calculate the temperature change (ΔT) for the water. Initial water temperature (C.3) – Final water temperature (C.3)	
B.7	1	Calculate the calories lost by the water. Heat (cal) lost by water = mass of water (C.5) $\times \Delta T$ (C.6) $\times$ specific heat (1.00 cal/g °C) This is the same number of calories that melted the ice.	
B.8	1	Calculate the grams of ice that melted. Final mass of the cup + water + melted ice (C.4) – initial mass of the cup + water (C.2)	
B.9	1	Calculate your experimental value for the heat of fusion for ice. Heat of fusion (cal/g) = $\dfrac{\text{heat (cal) gained to melt ice (C.7)}}{\text{Grams of ice (C.8)}}$	

C. Food Calories

Materials: Food products with nutrition data on labels

Section	Step	Procedure	Check when finished
C.1	1	Obtain a food product that has a Nutrition Facts label.	
C.1	2	Indicate the serving size.	

Section	Step	Procedure	Check when finished
C.2	1	List the grams of fat, carbohydrate, and protein in one serving of the food.	

Section	Step	Procedure	Check when finished
C.3	1	From the mass of each food type, calculate the Calories (kcal) of each food type in one serving using the accepted caloric values.	

Laboratory 7

Section	Step	Procedure	Check when finished
C.4	1	Determine the total Calories (kcal) in one serving.	

Section	Step	Procedure	Check when finished
C.5	1	Compare your total to the Calories listed on the upper portion of the label. Usually these totals are rounded to the nearest tens place.	

D. Exothermic and Endothermic Reactions

Materials: Two test tubes, test tube rack, water, scoop or spatula, $NH_4NO_3(s)$, anhydrous $CaCl_2(s)$, thermometer, stirring rod

Section	Step	Procedure	Check when finished
D.1	1	Place 5 mL of water in each of two test tubes.	
D.1	2	Record the temperature of the water.	
D.1	3	Add one scoop of $NH_4NO_3(s)$ crystals to the water in the first test tube. (See the first reaction below.)	
D.1	4	Add one scoop of *anhydrous* $CaCl_2(s)$ to the water in the second test tube. *Anhydrous* means "without water." (See the second reaction below.)	
D.1	5	Stir each with a stirring rod and record the temperature again.	

$$NH_4NO_3(s) \xrightarrow{H_2O} NH_4^+(aq) + NO_3^-(aq)$$

$$CaCl_2(s) \xrightarrow{H_2O} Ca^{2+}(aq) + 2\,Cl^-(aq)$$

Section	Step	Procedure	Check when finished
D.2	1	Describe each reaction as endothermic or exothermic.	
D.3	1	To each equation, add the term heat on the side of the reactants (if endothermic) and on the side of the products (if exothermic).	

Energy and Matter

Report Sheet - Lab 7

Date _____   Name _____
Section _____   Team _____
Instructor _____   _____

Pre-Lab Study Questions

1. Why is energy required for the boiling process?

2. When water at 0°C freezes, is heat lost or gained?

3. How is the caloric value (kcal/g) of a food sample determined?

4. How does an exothermic reaction differ from an endothermic reaction?

Energy and Matter

Report Sheet - Lab 7

Date _____ Name _____
Section _____ Team _____
Instructor _____

A. A Heating Curve for Water   Volume of water: _____ mL

A.1

Time (min)	Temperature (°C)	Time (min)	Temperature (°C)
0	_____	_____	_____
1	_____	_____	_____
2	_____	_____	_____
_____	_____	_____	_____
_____	_____	_____	_____
_____	_____	_____	_____
_____	_____	_____	_____
_____	_____	_____	_____
_____	_____	_____	_____

A.2 **Graphing the Heating Curve**

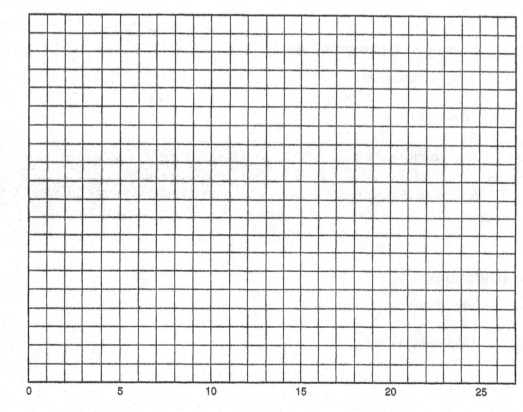

Laboratory 7

Report Sheet - Lab 7

A.3   Boiling point of water _____ °C
A.4   Temperature change (ΔT) _____ °C
A.5   Volume of water _____ mL
      Mass of water _____ g
A.6   Number of calories needed to heat water _____ cal
(Show calculations.)

Questions and Problems

Q.1   On the heating curve, how long did it take for the temperature to rise to 60°C?

B.   Energy in Changes of State

B.1   Empty calorimeter cup _____ g

B.2   Calorimeter + water _____ g

B.3   *Final* water temperature _____ °C

      Initial water temperature _____ °C

B.4   Calorimeter + water + melted ice _____ g

Calculations

B.5   Mass of water _____ g

B.6   Temperature change _____ °C

Energy and Matter

Report Sheet - Lab 7

C.7 Calories lost by water _____ cal
(Show calculations.)

Calories used to melt ice _____ cal

C.8 Mass of ice that melted _____ g
(Show calculations.)

C.9 Heat of fusion (calories to melt 1 g of ice) _____ cal/g
(Show calculations.)

Questions and Problems

Q.2 When water is heated, the temperature eventually reaches a constant value and forms a plateau on the graph. What does the plateau indicate?

Q.3 175 g of water was heated from 15° to 88°C. How many kilocalories were absorbed by the water?

Laboratory 7

Report Sheet - Lab 7

Q.4 How many calories are required at 0°C to melt an ice cube with a mass of 25 g?

Q.5   a.   Calculate the amount of heat (kcal) released when 50.0 g of water at 100°C hits the skin and cools to a body temperature of 37°C.

       b.   Calculate the amount of heat (kcal) released when 50.0 g of steam at 100°C hits the skin, condenses, and cools to a body temperature of 37°C.

       c.   Use your answer in 6a and 6b to explain why steam burns are so severe.

C.   Food Calories

C.1   Name of food product _____ Serving size _____

C.2   Mass of food types in one serving

       Carbohydrate _____ g   Fat _____ g   Protein _____ g

Energy and Matter

Report Sheet - Lab 7

C.3 Calculations for kcal per serving
(Show calculations.)
Carbohydrate _____ kcal (Cal)

Fat _____ kcal (Cal)

Protein _____ kcal (Cal)

C.4 Total Calories (kcal) per serving _____ kcal (Cal)

C.5 Calories (for one serving) listed on the label _____ Cal

Questions and Problems

Q.6 What percent (%) of the total Calories in your food product is from fat?

What percent (%) of the total Calories in your food product is from carbohydrate?

What percent (%) of the total Calories in your food product is from protein?

Q.7 How does your calculated number of Calories compare to the Calories listed on the label of the food product?

D. Exothermic and Endothermic Reactions

D.1 NH_4NO_3 $CaCl_2$

Initial temperature _____ _____

Final temperature _____ _____

Temperature change _____ _____

D.2 Endothermic or exothermic _____ _____

Laboratory 7

Report Sheet - Lab 7

A.3 Equations (add *heat*)

$$NH_4NO_3(s) \xrightarrow{H_2O} NH_4^+(aq) + NO_3^-(aq)$$

$$CaCl_2(s) \xrightarrow{H_2O} Ca^{2+}(aq) + 2\,Cl^-(aq)$$

Questions and Problems

Q.8 As a lab technician for a pharmaceutical company, you are responsible for preparing hot packs and cold packs. A hot pack involves the release of heat when a salt and water are mixed. A cold pack becomes colder because mixing a salt and water absorbs heat.

a. Which compound used in the experiment could you use to make a hot pack?

b. Which compound used in the experiment could you use to make a cold pack?

Q.9 When you burn a log in the fireplace or burn gasoline in a car, is the reaction (combustion) an endothermic or exothermic reaction?

8

Gas Laws: Boyle's and Charles'

Goals

- Graph the relationship between the pressure and volume of a gas.
- Observe the effect of changes in temperature upon the volume of a gas.
- Graph the data for volume and temperature of a gas.
- State a relationship between the temperature and volume of a gas.
- Use a graph to predict the value of absolute zero.

Text References

Review the following before performing this laboratory.
Timberlake; *General, Organic, and Biological Chemistry: Structures of Life*, Third Edition
Sections 7.1 to 7.6; Pages 262 to 277

or

Timberlake; *Chemistry: An Introduction to General, Organic, and Biological Chemistry*, Tenth Edition
Sections 6.1 to 6.6; Pages 210 to 226

Discussion

Pressure

The pressure of a gas depends upon the number of molecules hitting the walls of a container and their force. You increase pressure when you add air to a car or bicycle tire or basketball or when you blow up a balloon. Airplanes must be pressurized so that you breathe sufficient oxygen. Scuba divers require increased air pressures in their air tanks while diving because the pressure on their bodies increases. In the medical setting, blood pressure is measured to determine the force of the blood against the aorta and other blood vessels. In the operating room, an anesthesiologist monitors the pressures of oxygen and anesthesia given to a patient.

When we work with gases, we obtain the *atmospheric pressure* by reading a barometer in the laboratory. In a mercury barometer, air molecules strike the open surface of mercury, which pushes against the column of mercury in the vertical glass tube. When the height of the mercury column is 760 mm Hg, it measures a pressure of 1 atmosphere (atm). As the atmosphere pressure changes, the height of the mercury in the tube also changes.

Laboratory 8

A. Boyle's Law

According to Boyle's law, the pressure (P) of a gas varies inversely with the volume (V) of the gas when the temperature (T) and quantity, number of mole, (n) are kept constant. Mathematically, this is expressed as

$$P \sim \frac{1}{V} \quad (T, n \text{ constant}) \quad \text{Boyle's law}$$

By rearranging Boyle's law, we find that pressure × volume of a specific amount of gas is constant as long as the temperature and number of moles does not change.

$$P_1V_1 = \text{constant}$$

At the same temperature, the PV product of a gas remains constant. Therefore, we can write Boyle's law as an equality of the initial (*1*) pressure × volume with the final (*2*) pressure × volume.

$$P_1V_1 = P_2V_2 \; (T, n \text{ constant})$$

According to Boyle's law, volume decreases when the pressure increases, and volume increases when the pressure decreases. For example, if the pressure of a gas is doubled, the corresponding volume is one-half its previous value.

B. Charles' Law

In cold weather, the tires on a car or a bicycle seem to go flat. In hot weather, a full tire may burst. Such examples of changes in volume are related to changes in temperature. When temperature rises, the kinetic energy of the molecules increases. To keep pressure constant, the volume must expand. When the molecules slow down in cooler weather, the volume must decrease. According to Charles' law, the volume of a gas changes directly with the Kelvin temperature as long as the pressure and number of moles remain constant. Charles' law is expressed mathematically as

$$V \sim T_K \; (P \text{ and } n \text{ are constant})$$

$$\frac{V}{T_K} = constant$$

Under two different conditions, we can write Charles' law as follows:

$$\frac{V_1}{T_1} = \frac{V_2}{T_2}$$

Absolute Zero

In this experiment, the volume of a gas will be measured at different temperatures. The gas sample will be the amount of air contained in a 125-mL Erlenmeyer flask. A high temperature of about 100°C will be obtained by heating the flask and its air sample in boiling water. The lower temperatures for the gas sample will be obtained by placing the flask with the gas sample in pans of cool water, including one with ice water. Absolute zero is the theoretical value for the coldest temperature that matter can attain. Using the data for volume of a gas at different temperatures, a graph will be prepared that shows a relationship between decreasing volume and decreasing temperature. The value of absolute zero is predicted by extending the graph line (extrapolating) to the axis where the volume of the gas would decrease to 0 mL.

Lab Information

Time:	2 hr
Comments:	Be careful when you work around boiling water.
	Tear out the report sheets and place them beside the matching procedures.
Related Topics:	Pressure, volume, Boyle's law, Charles' law, combined gas law, Kelvin temperature

Experimental Procedures   *GOGGLES REQUIRED!*

A. Boyle's Law

Section	Step	Procedure	Check when finished
A.1	1	In an experiment, the volume of a specific amount of gas was measured at different volumes while the temperature was kept constant. You will work with the results given in the report sheet table.	
A.1	2	Determine the $P \times V$ product by multiplying the pressure and the volume in each sample.	
A.1	3	Round off the product to give the correct number of significant figures.	

Section	Step	Procedure	Check when finished
A.2	1	Review the graphing instructions on page xviii in the preface. Use as much of the graph area as possible.	
A.2	2	Mark the vertical axis in equal intervals of mm Hg of pressure. For the lowest pressure value, use a pressure that is slightly below the lowest value in the data: the highest value should be just above the highest measured value for pressure. For example, the pressure scale might begin at 600 mm Hg and go up to 1000 mm Hg.	
A.2	3	Divide the horizontal axis into equal intervals of mL. The volume values also should begin with a value near the smallest volume obtained.	

Laboratory 8

A.2	4	Plot the pressure (mm Hg) of the gas in each reading against the volume (mL).	
A.2	5	Draw a smooth line through the points obtained from the data. The data points will fall on a slight curve called a hyperbola, not a straight line.	
A.2	6	Use the graph to discuss the meaning of Boyle's law.	

B. Charles' Law

Materials: 125-mL Erlenmeyer flask, 400-mL beaker, one-hole rubber stopper with a short piece of glass tubing inserted and attached to a piece of rubber tubing, water containers, thermometer, pinch clamp, buret clamp, Bunsen burner (or a hot plate), wire gauze, iron ring and stand, a graduated cylinder, boiling chips, ice

Section	Step	Procedure	Check when finished
B	1	Dry any moisture on the inside of a 125-mL Erlenmeyer flask.	
B	2	Place a one-hole rubber stopper and tubing in the neck of the flask. (Be sure it has one hole.)	
B	3	Set a 400-mL beaker on a hot plate or on an iron ring covered by wire gauze. Add a few boiling chips to the bottom of the beaker.	
B	4	Attach a buret clamp to the neck of the flask and lower the flask into a 400-mL beaker without touching the flask to the bottom.	
B	5	Fasten the buret clamp to a ring stand.	
B	6	Pour water into the beaker until it comes up to the neck of the flask. Leave space at the top for the water to boil without boiling over. See Figure 8.1 for an illustration of one way set up a boiling water bath.	

Figure 8.1 Setup for heating an Erlenmeyer flask in a boiling water bath

Section	Step	Procedure	Check when finished
B.1	1	Begin heating and bring the water to a boil.	
B.1	2	Boil gently for 10 minutes to bring the temperature of the air in the flask to that of the boiling water.	
B.1	3	When you are ready to remove the flask, measure the temperature of the boiling water.	
B.1	4	Convert from degrees Celsius to the corresponding Kelvin temperature.	
B.1	5	Turn off the burner and immediately place a pinch clamp on the rubber tubing.	
B.1	6	Undo the buret clamp from the ring stand.	
B.1	7	In the lab, there should be 3 or 4 large containers with water at different temperatures. Using the buret clamp like a handle, *carefully* lift the flask out of the hot water and carry it over to one of the cool water containers. *Keeping the stopper end of the flask pointed downward,* immerse the flask in the cool water. See Figure 8.2.	
B.1	8	Keeping the flask *inverted*, remove the pinch clamp. Water will enter the flask as the air sample cools and decreases in volume.	

Figure 8.2 Place the heated flask in a pan of cool water.

Section	Step	Procedure	Check when finished
B.2	1	After the flask has been immersed in the cool water for at least 10 minutes, measure the temperature of the water in your cool water bath. We assume that the temperature of the cooled air sample in the flask is the same as the cool water outside the flask.	
B.2	2	Record.	
B.2	3	Convert the Celsius temperature to Kelvins.	

Laboratory 8

Section	Step	Procedure	Check when finished
B.3	1	*Keeping the flask inverted (upside down),* raise or lower the flask until the water level *inside* the flask is equal to the water level in the container. See Figure 8.3.	
B.3	2	While holding the flask at this level, reattach the pinch clamp to the rubber tubing.	
B.3	3	Remove the flask from the water, and set it upright on your desk.	
B.3	4	Remove the clamp.	
B.3	5	Use a graduated cylinder to measure the volume of water that entered the flask as the air sample cooled.	
B.3	6	Record the volume in mL.	

Figure 8.3 Equalize the water levels inside and outside the inverted flask.

Section	Step	Procedure	Check when finished
B.4	1	Fill the flask to the top with water but leave room for the volume occupied by the rubber stopper assembly.	
B.4	2	Measure the total volume of the flask by pouring the water into a graduated cylinder.	
B.4	3	Record.	

Section	Step	Procedure	Check when finished
B.5	1	List the temperatures (K) of the boiling water bath and the total gas volume (mL) obtained by other students in the lab.	
B.6	1	Calculate an *average* for the boiling water bath temperature (K).	
B.7	1	Calculate an *average* for the total gas volume (mL) of the flask.	
B.8	1	From the other students in the class, obtain the Kelvin temperatures for 4 other cool water baths and the volume of water that entered each of the other flasks. Use the *average* temperature for the temperature of the boiling water bath.	

B.8	2	Calculate the volume of air in the flask for each temperature by subtracting the amount of cool water that entered the flask from the average total gas volume. Volume of cooled air (B.8) = total flask volume (average) (B.7) – volume of water in flask (B.3)	
B.9	1	For each gas sample, calculate the V/T_K value for each. These should be constant, according to Charles' law. Any value that is not similar will not be a good data point to use on the graph.	
B.10	1	Graph the volume–temperature relationship of the gas. Review the graphing instructions on page xviii in the preface. The temperature (K) scale starts at 0 K and goes to 400 K. The volume axis extends from 0 mL to 150 mL. Theoretically, gases would reach a temperature called *absolute zero* at a volume of 0 mL.	
B.10	2	Draw the best straight line you can through the points you plotted on the graph. Then extend the line so it goes all the way to the temperature axis, which equals a volume of zero (0). Your prediction of absolute zero is the temperature where your extended line crosses the horizontal axis.	
B.10	3	Record this value. See Figure 8.4.	

Figure 8.4 Graph of temperature versus the volume of a gas

Report Sheet - Lab 8

Gas Laws: Boyle's and Charles'

Date _____ Name _____
Section _____ Team _____
Instructor _____ _____

Pre-Lab Study Questions

1. What are some occasions that require you to measure the pressure of a gas?

2. Why is an airplane pressurized?

3. Why does a scuba diver need increased gas pressure in the air tank?

4. How does temperature affect the kinetic energy of gas molecules?

Report Sheet - Lab 8

Date _____ Name _____
Section _____ Team _____
Instructor _____

A. Boyle's Law

A.1

Reading	Pressure (P)	Volume (V)	P × V (Product)
1	630 mm Hg	32.0 mL	
2	690 mm Hg	29.2 mL	
3	726 mm Hg	27.8 mL	
4	790 mm Hg	25.6 mL	
5	843 mm Hg	24.0 mL	
6	914 mm Hg	22.2 mL	

How do the values for the P × V product compare to each other?

Laboratory 8

Report Sheet - Lab 8

A.2   **Graphing Pressure and Volume: Boyle's Law**

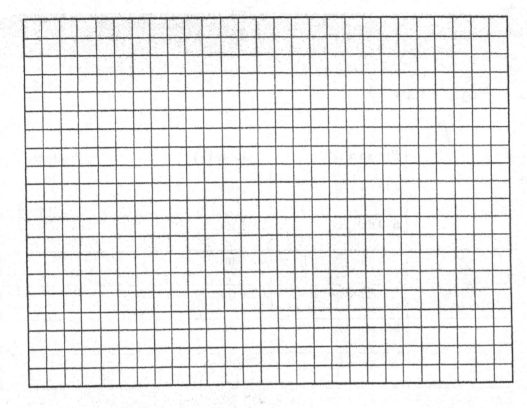

Questions and Problems

Q.1   According to your graph, what is the relationship between pressure and volume?

Q.2   On your graph, what is the volume of the gas at a pressure of 760 mm Hg?

Q.3   On your graph, what is the pressure of the gas when the volume is 30.0 mL?

Q.4   Complete the following when *T* and *n* remain constant:

Pressure	Volume
Increases	_____
_____	Increases
Decreases	_____

Report Sheet - Lab 8

Q.5   A sample of helium has a volume of 325 mL and a pressure of 655 mm Hg. What will be the pressure if the helium is compressed to 125 mL (*T* constant)? (*Show calculations.*)

Q.6   A 75.0-mL sample of oxygen has a pressure of 1.50 atm. What will be the new volume if the pressure becomes 4.50 atm (*T* constant)? (*Show calculations.*)

Q.7   In a weather report, the atmospheric pressure is given as 29.4 inches of mercury. What is the corresponding pressure in mm Hg? *(Show calculations.)*

Laboratory 8

Report Sheet - Lab 8

B. Charles' Law

B.1 Temperature of boiling water bath   _____ °C
                                           _____ K

B.2 Temperature of cool water   _____ °C
                                           _____ K

B.3 Volume of cool water that entered flask   _____ mL

B.4 Total gas volume of flask   _____ mL

B.5

Temperature of boiling water baths		Volume (mL) of flasks
°C	K	

B.6 Average boiling water bath temperature (K) _____ K

B.7 Average total gas volume in flask   _____ mL

Report Sheet - Lab 8

B.8

Water bath	Temperature of water bath		Average total gas volume	Cool water that entered flask	Volume of air remaining in flask
	°C	K			
Boiling water bath			0.0 mL		
Cool bath 1					
Cool bath 2					
Cool bath 3					
Cool bath 4					
Cool bath 5					

B.9

	Boiling water bath	Bath 1	Bath 2	Bath 3	Bath 4	Bath 5
Volume (mL)						
Temperature (K)						
$\dfrac{Volume}{Temperature}$						

Laboratory 8

Report Sheet - Lab 8

B.10 Graphing Volume and Temperature Relationship

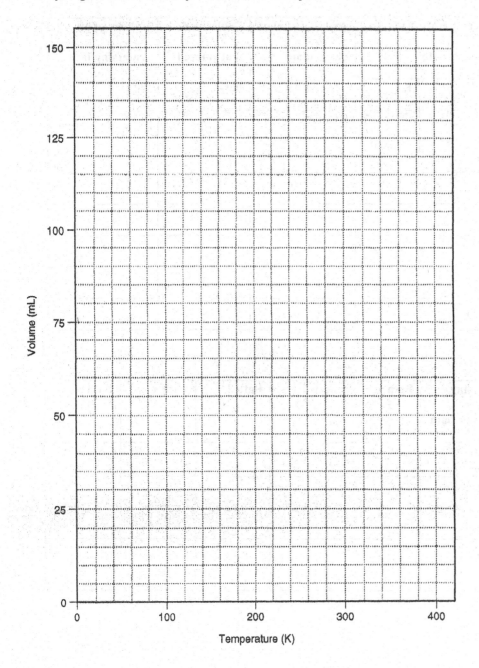

According to your graph, what is the predicted Kelvin temperature of absolute zero?

Gas Laws: Boyle's and Charles'

Report Sheet - Lab 8

How does your predicted value for absolute temperature compare with the accepted value of 0 K?

Questions and Problems

Q.8   Using your graph, how are temperature and volume related?

Q.9   Indicate the change expected when pressure is held constant for a given amount of gas.

Volume   Temperature

Increases   _____

_____   Decreases

Decreases   _____

_____   Increases

Q.10   A gas with a volume of 525 mL at a temperature of –25°C is heated to 175°C. What is the new volume of the gas if pressure and number of moles are held constant?

Laboratory 8

Report Sheet - Lab 8

Q.11　A gas has a volume of 2.8 L at a temperature of 27°C. What temperature (°C) is needed to expand the volume to 15 L? (*P* and *n* are constant.)

Q.12　Combined gas law problem: A balloon is filled with 500.0 mL of helium at a temperature of 27°C and 755 mm Hg. As the balloon rises in the atmosphere, the pressure and temperature drop. What volume will it have when it reaches an altitude where the temperature is –33°C and the pressure is 0.65 atm?

9

Partial Pressures of Oxygen, Nitrogen, and Carbon Dioxide

Goals
- Measure the percentage of O_2 and N_2 in the air.
- Determine the partial pressures of O_2 and N_2 in air.
- Determine the partial pressure of CO_2 in air.
- Determine the partial pressure of CO_2 in expired (exhaled) air.

Text References

Review the following before performing this laboratory.

Timberlake; *General, Organic, and Biological Chemistry: Structures of Life*, Third Edition
Sections 9.1 to 9.6; Pages 337 to 365

or

Timberlake; *Chemistry: An Introduction to General, Organic, and Biological Chemistry*, Tenth Edition
Section 5.9; Pages 195 to 198

Discussion

A. Partial Pressures of Oxygen and Nitrogen in Air

The two major gases in the air are oxygen (O_2) and nitrogen (N_2). In this experiment, you will make measurements to use in calculating the percentage of nitrogen and oxygen in the air. Removing the oxygen from a sample of air and measuring the change in volume will determine the amount of oxygen in air. To remove the oxygen, we place some iron filings in a moistened test tube and allow them to react with the oxygen.

$$4Fe + 3O_2 \rightarrow 2Fe_2O_3$$

As the oxygen reacts, the volume it occupied in the air sample is replaced by water. By measuring the volume of the air in the container, and the remaining volume of nitrogen, the volume of oxygen and its percentage in air can be calculated. Using the atmospheric pressure, the partial pressures (mm Hg) of oxygen gas and nitrogen gas in the air can be calculated. This assumes that all other gases in air are negligible for this experiment. Partial pressures are the individual pressures exerted by each of the gases that make up the total atmospheric pressure.

Laboratory 9

B., C. Carbon Dioxide in the Atmosphere and Expired Air

When the body metabolizes nutrients, one of the end products is carbon dioxide, CO_2. The level of carbon dioxide in the blood triggers breathing mechanisms and provides for the correct pH of the blood. Accumulation of carbon dioxide above these levels can result in respiratory and metabolic dysfunction and possible death. The body eliminates most of the carbon dioxide by exhalation of air from the lungs. Plants utilize most of the CO_2 in the atmosphere and return O_2 to the atmosphere.

In experiments B and C, the partial pressures of carbon dioxide in the atmosphere and in expired (exhaled) air will be determined by reacting CO_2 with NaOH.

$$CO_2(g) + NaOH \rightarrow NaHCO_3$$

The other major gases in the atmosphere, oxygen and nitrogen, do not react with sodium hydroxide and continue to exert their respective partial pressures.

Lab Information

Time: 2 hr (part A is finished on the next lab day)
Comments: Tear out the report sheets and place them beside the matching procedures.
Related Topics: Partial pressure, Dalton's law, gas mixtures, atmospheric pressure

Dispose of all chemicals as directed by your lab instructor.

Experimental Procedures *GOGGLES REQUIRED!*

A. Partial Pressures of Oxygen and Nitrogen in Air

Materials: 250-mL beaker, large test tube, iron filings, graduated cylinder, clamp or test tube holder

Section	Step	Procedure	Check when finished
A.1	1	Completely fill a large test tube with water.	
A.1	2	Measure the volume of the water in the test tube by emptying the water into a graduated cylinder. This is equal to the volume of air in the test tube.	
A.1	3	Obtain a small scoop of iron filings and sprinkle them in the empty but moist test tube.	
A.1	4	Shake the iron filings about the test tube. Some should adhere to the sides. Shake out any loose filings in the test tube.	
A.1	5	Fill a 250-mL beaker about one-half full of water.	
A.1	6	Place the test tube containing the iron filings upside down in the water. See Figure 9.1	

Partial Pressures of Oxygen, Nitrogen, and Carbon Dioxide

			Check when finished
A.1	7	Attach a clamp or test tube holder to the test tube and let the handle rest on the rim of the beaker. This will stabilize the inverted test tube.	
A.1	8	Carefully store the beaker and test tube in a place where they will be undisturbed. This part of the experiment will be finished at the beginning of the next laboratory period. See Figure 11.1.	

Figure 9.1 Beaker and test tube assembly for determination of the partial pressures of oxygen and nitrogen in the air

Section	Step	Procedure	Check when finished
A.2	1	***Next laboratory period*** At your next laboratory class, you should find that water has filled part of the test tube. This happened because the iron filings in the test tube reacted with the oxygen in the air. As the oxygen is used up, it is replaced by water.	
A.2	2	Use a marking pen or tape to mark the water level inside the test tube.	
A.2	3	Remove the test tube from the beaker.	
A.2	4	Fill the test tube with water *up to the line* you marked.	
A.2	5	Empty the water into a graduated cylinder	
A.2	6	Record the volume. This volume is equal to the volume of the nitrogen in the air that *remains* in the test tube after the oxygen reacted.	

119

Laboratory 9

Calculations

Section	Step	Procedure	Check when finished
A.3	1	Read a barometer and record the atmospheric pressure (mm Hg).	
A.4	1	Calculate the volume (mL) of oxygen initially present in the test tube by subtracting the volume (mL) of nitrogen gas from the total volume of the test tube. Volume (O_2) = Volume of test tube (A.1) – Volume of nitrogen (N_2) (A.2)	
A.5	1	Calculate the percentages of oxygen and nitrogen in air by dividing the O_2 (or N_2) volume by the total volume of the initial air sample. $\dfrac{Volume\ gas\ (A.4\ or\ A.2)}{Volume\ of\ Air\ (A.1)} \times 100 = \%gas\ in\ air$	
A.6	1	Calculate the partial pressures of oxygen (O_2) and nitrogen (N_2). Multiply the atmospheric pressure by the percentage you calculated for each gas. $\dfrac{\%\ gas\ (A.5)}{100} \times atmospheric\ pressure = partial\ pressure\ of\ gas$	

B. Carbon Dioxide in the Atmosphere (*This may be a demonstration by the instructor.*)

Materials: 250-mL Erlenmeyer flask to fit two-hole stopper, shell vial, 6 M NaOH, mineral oil, 150-mL beaker, food coloring (optional), meterstick, glass tubing (60–75 cm), two-hole stopper with two short pieces of glass tubing, rubber tubing (1 short, 1 long), funnel, pinch clamp

Section	Step	Procedure	Check when finished
B.1	1	Read a barometer	
B.1	2	Record the atmospheric pressure (mm Hg).	

Section	Step	Procedure	Check when finished
B.2	1	Carefully lower an empty shell vial into a 250-mL Erlenmeyer flask so that the vial remains upright. See Figure 9.2.	
B.2	2	Set a funnel in the flask directly above the shell vial.	
B.2	3	Obtain a small amount of 6 M NaOH in a small beaker. ***Caution: Sodium hydroxide is caustic! Be sure to clean up any spill immediately. Wash any spills on skin for 10 minutes.***	

Partial Pressures of Oxygen, Nitrogen, and Carbon Dioxide

B.2	4	*Slowly* pour the NaOH through the funnel into the vial and stop when the vial is about 3/4 full.	
B.2	5	Pour a small amount of mineral oil into the vial until the oil forms a *thin* layer of about 1 mm on top of the NaOH. The mineral oil on top of the sodium hydroxide prevents it from reacting with carbon dioxide.	
B.2	6	Remove the funnel.	
B.2	7	Prepare a setup as shown in Figure 9.3 and described below.	
B.2	8	Place the two-holed stopper containing two pieces of glass tubing in the flask. Make sure that the stopper is tight.	
B.2	9	Attach a short piece of rubber tubing (A) to one piece of glass tubing.	
B.2	10	Attach a longer piece of rubber tubing to the other glass tubing in the stopper (B).	
B.2	11	To the open end of the long piece of rubber tubing, attach a long piece (60–70 cm) of glass tubing.	
B.2	12	Place the open end of the glass tubing in a beaker containing water and a few drops of food coloring to make it easier to see the level of the water (optional).	
B.2	13	With the rubber tubing open, the flask is full of air at atmospheric pressure. *Close* the system by attaching a pinch clamp to the short piece of rubber tubing (A).	

Figure 9.2 Filling a shell vial with NaOH using a funnel in a flask

Laboratory 9

Figure 9.3 Flask, tubing, and vial setup for carbon dioxide determination

Section	Step	Procedure	Check when finished
B.2	14	Gently tilt the flask and tip over the shell vial.	
B.2	15	The NaOH solution will spill out into the bottom of the flask.	
B.2	16	Swirl or shake the flask gently. (Hold the flask around the top with your fingers. Try not to let your hands warm the flask, because an increase in temperature will change the pressure.) *Make sure that the end of tube B stays below the water level in the beaker.*	
B.2	17	As the carbon dioxide in the sample reacts with NaOH, there is a drop in the pressure of the gases in the sample, and water will rise in the glass tube. *However, because there is only a small amount of CO_2 in the air, the water level will not change very much.* $$CO_2(g) + NaOH \rightarrow NaHCO_3$$	
B.2	18	Continue to swirl the flask until there is no further change in the water level. *Make sure that the end of tube B remains below the water level in the beaker.*	
B.2	19	Using a meterstick, measure the distance in millimeters (mm) between the water level in the beaker and the water level in the vertical glass tube B. This is equal to the partial pressure of CO_2 in mm H_2O.	

Calculations

Section	Step	Procedure	Check when finished
B.3	1	Calculate the P_{CO_2} in mm Hg by dividing the P_{CO_2} (mm H_2O) by 13.6. (1.00 mm Hg exerts the same pressure as 13.6 mm H_2O.) This is equal to the partial pressure of CO_2 (mm Hg). $1 \text{ mm Hg} = 13.6 \text{ mm } H_2O$ $\text{Height in mm } H_2O \text{ (B.2)} \times \dfrac{1 \text{ mm Hg}}{13.6 \text{ mm } H_2O} = P_{CO_2} \text{ in the atmosphere (mm Hg)}$	
B.4	1	Calculate the percent CO_2 in the atmosphere by dividing the P_{CO_2} (B.3) by the atmospheric pressure. $\dfrac{P_{CO_2} \text{ (mm Hg)(B.3)}}{P \text{ atmosphere (mm Hg)(B.1)}} \times 100 = \% \; CO_2 \text{ in atmosphere}$	

C. Carbon Dioxide in Expired Air

Materials: 250-mL Erlenmeyer flask to fit two-hole stopper, shell vial, 6 M NaOH, mineral oil, 150-mL beaker, food coloring (optional), meterstick, glass tubing (60–75 cm), two-hole stopper with two short pieces of glass tubing, rubber tubing (1 short, 1 long), funnel, pinch clamp, clean straws. *Caution: NaOH is caustic, clean up any spills immediately.*

Section	Step	Procedure	Check when finished
C.1	1	Read a barometer.	
C.1	2	Record the atmospheric pressure (mm Hg).	

Section	Step	Procedure	Check when finished
C.2	1	Rinse out the flask you used in part B.	
C.2	2	Set up the apparatus as you did in part B. This time, place a clean straw into the rubber tubing A. Place glass tube B in the beaker of water.	
C.2	3	Take a breath of air, hold for a moment, and exhale through the straw. This will cause bubbling in the beaker of water. **Caution: Take your time. Exhaling too rapidly may cause hyperventilation and make you dizzy. Stop and rest.**	
C.2	4	Cover the straw with your finger while you inhale. Exhale through the straw again.	

Laboratory 9

Section	Step	Procedure	Check when finished
C.2	5	Repeat this process 3 times.	
C.2	6	Place a pinch clamp on the short piece of rubber tubing (A).	
C.2	7	Remove the straw. Now the flask is filled with expired (exhaled) air.	
C.2	8	Gently tilt the flask and tip over the shell vial, allowing the NaOH to spill out into the bottom of the flask.	
C.2	9	Swirl or shake the flask gently. (Hold the flask around the top with your fingers. Try not to let your hands warm the flask, because an increase in temperature will change the pressure.) *Make sure that the end of tube B remains below the water level in the beaker.* As the carbon dioxide in the sample reacts with NaOH, there is a drop in the pressure of the gases in the sample, and water will rise in the glass tube. Because of the higher partial pressure of CO_2 in exhaled air, the water level in tube B will make a dramatic rise. $$CO_2(g) + NaOH \rightarrow NaHCO_3$$	
C.2	10	Keep swirling the flask a few minutes until there is no further change in the water level.	
C.2	11	Using a meterstick, measure the distance (mm) *between* the water level in the beaker and the water level in the vertical glass tube. Make sure your measurement is in millimeters (mm), not centimeters (cm). This is the partial pressure of CO_2 in mm H_2O as done previously.	

Calculations

Section	Step	Procedure	Check when finished
C.3	1	Calculate the P_{CO_2} in mm Hg by dividing the P_{CO_2} (mm H_2O) by 13.6. (1 mm Hg exerts the same pressure as 13.6 mm H_2O.) This is equal to the partial pressure of CO_2 (mm Hg) for expired air. $$1 \text{ mm Hg} = 13.6 \text{ mm } H_2O$$ $$\text{Height in mm } H_2O \text{ (C.2)} \times \frac{1 \text{ mm Hg}}{13.6 \text{ mm } H_2O} \times 100 = \% \ CO_2 \text{ in expired air}$$	
C.4	2	Calculate the percent CO_2 in expired air by dividing the P_{CO_2} (C.3) by the atmospheric pressure (C.1). $$\frac{P_{CO_2} \text{ (mm Hg)(C.3)}}{P \text{ atmosphere (mm Hg)(C.1)}} \times 100 = \% \ CO_2 \text{ in expired air}$$	

Partial Pressures of Oxygen, Nitrogen, and Carbon Dioxide

Report Sheet - Lab 9

Date _____ Name _____
Section _____ Team _____
Instructor _____

Pre-Lab Study Questions

1. What is meant by the term *partial pressure*?

2. Do you predict there will be more CO_2 in exhaled gas than in the atmosphere. Why or why not.

3. How will you determine the partial pressure of N_2 in the atmosphere when it does not react in any of the experiments?

Laboratory 9

Partial Pressures of Oxygen, Nitrogen, and Carbon Dioxide

Report Sheet - Lab 9

Date _____ Name _____
Section _____ Team _____
Instructor _____   _____

A. Partial Pressures of Oxygen and Nitrogen in Air

A.1 Volume of test tube _____ mL

A.2 Volume of nitrogen _____ mL

A.3 Atmospheric pressure _____ mm Hg

Calculations

A.4 Volume of oxygen _____ mL

A.5 Percent oxygen _____ % O_2
(Show calculations.)

Percent nitrogen _____ % N_2
(Show calculations.)

A.6 Partial pressure of oxygen _____ mm Hg

Partial pressure of nitrogen _____ mm Hg

Questions and Problems

Q.1 If you lived in the mountains where the atmospheric pressure is 685 mm Hg, what would the partial pressure of oxygen and of nitrogen be? Use your % values from A.5.

Laboratory 9

Report Sheet - Lab 9

B. Carbon Dioxide in the Atmosphere

B.1   Atmospheric pressure   _____ mm Hg

B.2 P_{CO_2} (height of water column in tube B)   _____ mm H_2O

Calculations

B.3 P_{CO_2}

$$\underline{} \text{ mm } H_2O \times \frac{1 \text{ mm Hg}}{13.6 \text{ mm } H_2O} = \underline{} \text{ mm Hg}$$

B.4   Percent CO_2 in the atmosphere   _____ % CO_2 (atmosphere)
      (Show calculations.)

C. Carbon Dioxide in Expired Air

C.1   Atmospheric pressure   _____ mm Hg

C.2   Height of water column   _____ mm H_2O

Calculations

C.3 P_{CO_2}

$$\underline{} \text{ mm } H_2O \times \frac{1 \text{ mm Hg}}{13.6 \text{ mm } H_2O} = \underline{} \text{ mm Hg}$$

C.4   Percent CO_2 (expired air)   _____ % CO_2 (expired air)
      (Show calculations.)

Questions and Problems

Q.2   Which would you expect to be higher, the percentage of CO_2 in the atmosphere or in expired air? Why?

Q.3   What is the total pressure in mm Hg of a sample of gas that contains the following gases: O_2 (45 mm Hg), N_2 (1.20 atm), and He (825 mm Hg)?

Q.4   A mixture of gases has a total pressure of 1650 mm Hg. The gases in the mixture are helium (215 mm Hg), nitrogen (0.28 atm), and oxygen. What is the partial pressure of the oxygen in atm?

10

Solutions

Goals

- Observe the solubility of a solute in polar and nonpolar solvents.
- Determine the effect of particle size, stirring, and temperature on the rate of solution formation.
- Identify an unsaturated and a saturated solution.
- Observe the effect of temperature on solubility.
- Measure the solubility of KNO_3 at various temperatures, and graph a solubility curve.
- Calculate the mass/mass percent and mass/volume percent concentrations for a NaCl solution.
- Calculate the molar concentration of the NaCl solution.

Text References

Review the following before performing this laboratory.

Timberlake; *General, Organic, and Biological Chemistry: Structures of Life*, Third Edition
Sections 8.1 and 8.3 to 8.5; Pages 296 to 299 and 305 to 320

or

Timberlake; *Chemistry: An Introduction to General, Organic, and Biological Chemistry*, Tenth Edition
Sections 7.1 and 7.3 to 7.5; Pages 242 to 250 and 251 to 264

Discussion

A. Polarity of Solutes and Solvents

A solution is a homogenous mixture of the particles of two or more substances. The substance that is present in the greater amount is called the *solvent*. The substance that is present in the smaller amount is the *solute*. In many solutions, including body fluids and the oceans, water is the solvent. Water is considered the *universal solvent*. However, the solutes and solvents that make up solutions may be solids, liquids, or gases. Carbonated beverages are solutions of CO_2 gas in water.

    A solution forms when the attractive forces between the solute and the solvent are similar. A polar (or ionic) solute such as NaCl is soluble in water, a polar solvent. As the NaCl dissolves, its ions separate into Na^+ and Cl^-. The positive Na^+ ions are attracted to the partially negative oxygen atoms of water. At the same time, the negative Cl^- ions are pulled into the solution by their attraction to the partially positive hydrogen atoms of water. Once the ions are into the solution, they stay in solution because they are hydrated, which means that a group of water molecules is attracted to each ion.

    Water, which is polar, dissolves polar solutes such as glucose and salt, NaCl. A weakly nonpolar solvent such as acetone is needed to dissolve a nonpolar solute such as nail polish. This requirement of similar electrical attraction between solute and solvent is sometimes stated as "like dissolves like."

Laboratory 10

B. Solubility of KNO$_3$

When a solution holds the maximum amount of solute at a certain temperature, it is saturated. When more solute is added, the excess appears as a solid in the container. The maximum amount of solute that dissolves is called the solubility of that solute in that solvent. Solubility is usually stated as the number of grams of solute that dissolve in 100 mL (or 100 g) of water. The solubility depends upon several factors, including the nature of the solute and solvent, the temperature, and the pressure (for a gas). Most solids are more soluble in water at higher temperatures. Generally, the dissolving of a solid solute is endothermic, which means that solubility increases with an increase in temperature.

C. Concentration of a Sodium Chloride Solution

The concentration of a solution is calculated from the amount of solute present in a certain amount of solution. The concentration may be expressed using different units for amount of solute and solution. A *mass/mass percent (m/m)* concentration expresses the grams of solute in the grams of solution. The *mass/volume (m/v) percent* concentration of a solution states the grams of solute present in the milliliters of the solution.

$$\text{mass/mass percent (m/m)} = \frac{\text{grams of solute}}{\text{grams of solution}} \times 100$$

$$\text{mass/volume percent (m/v)} = \frac{\text{grams of solute}}{\text{milliliters of solution}} \times 100$$

A *molar* (M) concentration gives the moles of solute in a liter of solution.

$$\text{molarity (M)} = \frac{\text{moles of solute}}{1 \text{ liter of solution}}$$

In this experiment, you will measure a 10.0-mL volume of a sodium chloride solution. The mass of the solution will be determined by weighing the solution in a preweighed evaporating dish. After the sample is evaporated to dryness, it is weighed again. From this data, the mass of the salt (solute) is obtained. Using the mass of the solute and the mass of the solution, the mass/mass (m/m) percent can be calculated. From the mass of the solute and the volume (mL) of the solution, the mass/volume (m/v) percent can be calculated. To calculate the molarity of the solution, convert the mass of the solute to moles, and the volume to liters (L).

Lab Information

Time: 3 hr

Comments: Some solvents in part A of the experiments are flammable. Do not light any burners. Tear out the report sheets and place them beside the matching procedures. Your instructor will use the conductivity apparatus in a demonstration for part B.

Related Topics: Solute, solvent, formation of solutions, polar and nonpolar solutes, electrolytes, nonelectrolytes, concentrations of solutions

Experimental Procedures GOGGLES REQUIRED!

A. Polarity of Solutes and Solvents (*This may be a demonstration by your instructor.*)

Materials: Test tubes (8), test tube rack, spatulas, stirring rods, $KMnO_4(s)$, $I_2(s)$, sucrose(s), vegetable oil, cyclohexane

Section	Step	Procedure	Check when finished
A.1	1	**Solubility of solutes in a polar solvent** Set up four test tubes in a test tube rack.	
A.1	2	To each test tube, add a few crystals (or a few drops) of a solute: $KMnO_4$, I_2, sucrose, or vegetable oil. *Iodine (I_2) can burn the skin. Handle cautiously!*	
A.1	3	To each, add 3 mL of water and stir the mixture with a glass stirring rod.	
A.1	4	Describe each solute as soluble or not soluble in the water, a polar solvent.	
A.1	5	Save for comparison to the test tubes in A.2.	

Section	Step	Procedure	Check when finished
A.2	1	**Solubility of solutes in a nonpolar solvent** To a different set of four test tubes, add a few crystals (or a few drops) of a solute: $KMnO_4$, I_2, sucrose, or vegetable oil.	
A.2	2	Place 3 mL of cyclohexane, a nonpolar solvent, in each. *Caution: Cyclohexane is <u>flammable</u>—do not proceed if any laboratory burners are in use. Use in fume hood if possible.*	
A.2	3	Indicate whether each solute is soluble or not soluble in cyclohexane, a nonpolar solvent	
A.2	4	Compare the solubility of the solutes in cyclohexane (A.2) with their solubility in water (A.1).	
A.2	5	Discard the solutions for A.1 and A.2 in the waste containers provided in the lab, *NOT* in the sink.	

Laboratory 10

Section	Step	Procedure	Check when finished
A.3	1	Determine the polarity of each solute from its solubility in each type of solvent. If a solute dissolves in a polar solvent like water, it is a polar solute. If a solute dissolves in a nonpolar solvent like cyclohexane, the solute is nonpolar.	

B. Solubility of KNO_3

Materials: Weighing paper or small container, spatula, stirring rod, large test tube, 400-mL beaker, buret clamp, hot plate or Bunsen burner, thermometer, 10-mL graduated cylinder, $KNO_3(s)$

To reduce the amount of KNO_3 used, each group of students will be assigned an amount of KNO_3 to weigh out. The results will be shared with the class.

Section	Step	Procedure	Check when finished
B.1	1	Obtain a piece of weighing paper or a small container and weigh it carefully. (Or you may tare the container.)	
B.2	1	Each group of students will be assigned an amount of KNO_3 from 2 to 7 grams. Weigh out an amount of KNO_3 that is close to your assigned amount. For example, if you are assigned an amount of 3 grams, measure out a mass such as 3.10 g or 3.25 g or 2.85 g. It is not necessary to add or remove KNO_3 to obtain exactly 3.00 g. Weigh carefully.	
B.2	2	Calculate the mass of KNO_3.	
B.2	3	***Taring a container on an electronic balance:*** The mass of a container on an electronic balance can be set to 0 by pressing the tare bar. As a substance is added to the container, the mass shown on the readout is for the substance only. (When a container is tared, it is not necessary to subtract the mass of the beaker.)	

Section	Step	Procedure	Check when finished
B.3	1	The temperature at which the KNO_3 is soluble is determined by heating and cooling the KNO_3 solution. Place 5.0 mL of water in a large test tube.	
B.3	2	Add your weighed amount of KNO_3.	
B.3	3	Clamp the test tube to a ring stand and place the test tube in a beaker of water. See Figure 10.1.	

Section	Step	Procedure	Check when finished		
B.3	4	Use a hot plate or Bunsen burner to heat the water. See Figure 13.1.			
B.3	5	Stir the mixture and continue heating until all the KNO_3 dissolves.			
B.3	6	As soon as all the KNO_3 dissolves, turn off the burner or hot plate.			
B.3	7	Loosen the clamp and remove the test tube from the hot water.			
B.3	8	As the test tube and contents cool, stir gently with a thermometer.			
B.3	9	Look closely for the first appearance of crystals. As soon as you see some solid crystals, read the temperature of the solution.			
B.3	10	Record. This is the temperature at which the solution becomes saturated. The amount of KNO_3 in that solution is the solubility of KNO_3 at that temperature.			
B.3	11	Repeat the warming and cooling (steps 3 through 10) of the solution until you have obtained three or more temperature readings that agree.			
B.3	12	Set the test tube aside.			
B.3	13	In 15–20 minutes, observe the appearance of the crystals in the test tube.			
B.3	14	To discard, add water and heat until the KNO_3 dissolves.			
B.3	15	Pour the solution in proper waste containers provided in the laboratory, NOT in the sink. (Solid KNO_3 can be recovered from the solution by evaporation to dryness.)			

Figure 10.1 Heating the KNO_3 solution in a water bath

Calculations

Section	Step	Procedure	Check when finished
B.4	1	Solubility is expressed as the number of grams of solute in	

Laboratory 10

		100 mL of water. Because you used a sample of 5.0 mL of water, the mass of the solute you measured out and the 5.0 mL of water are both multiplied by 20. $$\frac{g\,KNO_3}{5.0\,mL\,water} \times \frac{20}{20} = \frac{g\,KNO_3}{100\,mL\,water} = Solubility\,(g\,KNO_3\,per\,100\,mL\,water)$$	
B.4	2	Collect the solubility results of other KNO_3 solutions and their solubility temperatures from the other groups of students in the lab.	
B.5	1	Prepare a graph of the solubility curve for KNO_3. Plot the solubility (g KNO_3/100 mL water) on the vertical axis and the temperature (0–100°C) on the horizontal axis.	

C. Concentration of a Sodium Chloride Solution

Materials: Hot plate (or Bunsen burner, iron ring, and wire screen), evaporating dish, NaCl solution, 400-mL beaker (to fit evaporating dish), 10-mL graduated cylinder (or 10-mL pipet)

Section	Step	Procedure	Check when finished
C.1	1	Weigh a dry evaporating dish.	
C.1	2	Record the mass. *Do not round off.*	

Section	Step	Procedure	Check when finished
C.2	1	Using a 10.0-mL graduated cylinder, or a 10.0-mL pipet, measure out a 10.0-mL sample of the NaCl solution. See Figure 10.2.	
C.2	2	Record this volume.	
C.2	3	*Using a pipet:* Place the pipet bulb on the upper end of the pipet.	
C.2	4	Squeeze the bulb about halfway.	
C.2	5	Place the tapered end of the pipet in the liquid.	
C.2	6	Release the pressure on the bulb. As the bulb inflates, liquid will move into the pipet.	
C.2	7	The level of liquid should rise above the volume mark, but not into the bulb.	
C.2	8	Remove the bulb and quickly cover the pipet with your *index* finger. Wipe off the outside of the pipet with a paper towel.	
C.2	9	Adjust the pressure of your finger to slowly drain the liquid until the level is at the volume mark. You may need to practice.	

Section	Step	Procedure	Check when finished
C.2	10	With your finger still on the pipet, lift the pipet out of the liquid and move it to the evaporating dish.	
C.2	11	Lift your finger off the pipet to let the liquid flow out. Some liquid should remain in the tip.	

Figure 10.2 Using a pipet

Section	Step	Procedure	Check when finished
C.3	1	Weigh the evaporating dish and the NaCl solution.	
C.3	2	Record.	

Section	Step	Procedure	Check when finished
C.4	1	Fill a 400-mL beaker about half full of water.	
C.4	2	Set on a hot plate, or heat with a Bunsen burner using an iron ring with a wire screen. See Figure 10.3.	
C.4	3	Place the evaporating dish on top of the beaker	
C.4	4	Heat the water in the beaker to boiling. You may need to add more water to the bath as you proceed.	
C.4	5	When the NaCl appears to be dry or begins to pop, turn off the burner.	
C.4	6	After the evaporating dish has cooled, dry the bottom and place it directly on the hot plate or in the iron ring.	
C.4	7	Heat gently with a low flame to dry the salt completely.	
C.4	8	Allow the evaporating dish and dried NaCl sample to cool.	
C.4	9	Weigh the evaporating dish and the dry NaCl.	
C.4	10	Record. *Do not round off.*	

Laboratory 10

Figure 10.3 Using a Bunsen burner or hot plate to evaporate a salt solution

Calculations

Section	Step	Procedure	Check when finished
C.5	1	Calculate the mass of the solution. (D.3) − (D.1)	
C.6	1	Calculate the mass of the NaCl after drying. Subtract the mass of the evaporating dish (D.1) from the total mass of the evaporating dish and the dried salt (D.4).	
C.7	1	Calculate the mass/mass percent concentration using the mass of the dry NaCl (D.6) divided by the mass of the solution (D.5). $$\text{Mass/mass percent} = \frac{\text{Mass of dry NaCl}}{\text{mass (g) of solution}} \times 100$$	
C.8	1	Calculate the mass/volume percent concentration by dividing the mass of the dry NaCl (D.6) by the volume of the NaCl solution (D.2) and multiplying by 100. $$\text{Mass/volume percent} = \frac{\text{Mass of dry NaCl}}{\text{volume (mL) of solution}} \times 100$$	
C.9	1	Calculate the moles of NaCl by dividing the grams of dried NaCl (D.6) divided by the molecular weight of NaCl. The molar mass of NaCl is 58.5 g/mole. $$\text{g of dried NaCl} = \frac{1 \text{ mole NaCl}}{58.5 \text{ g NaCl}} \times \text{moles NaCl}$$	
C.10	1	Convert the volume in mL of the solution to the corresponding volume in liters by dividing the volume of NaCl (D.2) by 1000. $$\text{mL of NaCl solution} \times \frac{1 \text{ L}}{1000 \text{ mL}} = \text{L of NaCl solution}$$	
C.11	1	Calculate the molarity of the NaCl solution by dividing the moles of NaCl (D.9) by the liters of NaCl solution (D.10). $$\text{Molarity (M)} = \frac{\text{moles NaCl}}{\text{L of solution}}$$	

Report Sheet - Lab 10

Date _____   Name _____
Section _____   Team _____
Instructor _____

Pre-Lab Study Questions

1. Why does an oil-and-vinegar salad dressing have two separate layers?

2. What is meant by the mass/mass percent concentration of a solution?

3. Are salts more soluble in water at higher temperature or lower temperature?

4. What is molarity?

Solutions

Report Sheet - Lab 10

Date _____   Name _____
Section _____   Team _____
Instructor _____

A. Polarity of Solutes and Solvents

Solute	Soluble/Not Soluble in		Identify the Solute as Polar or Nonpolar
	Water (A.1)	Cyclohexane (A.2)	(A.3)
$KMnO_4$			
I_2			
Sucrose			
Vegetable oil			

Questions and Problems

Q.1   NaCl is soluble in water, but I_2 is not. Explain.

B. Solubility of KNO_3

B.1 Mass of Container	B.2 Mass of Container + KNO_3	Mass of KNO_3	B.3 Temperature (Crystals Appear)	B.4 Solubility (g KNO_3/100 mL H_2O)

Laboratory 10

Report Sheet - Lab 10

B.5 Graphing the solubility of KNO_3 vs. temperature (°C)

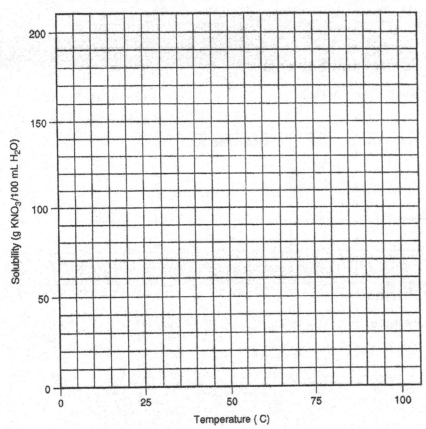

According to your graph, what is the effect of increasing temperature on the solubility of KNO_3?

Questions and Problems

Q.2 The solubility of sucrose (common table sugar) at 70°C is 320 g/100 g H_2O.

a. How much sucrose can dissolve in 200 g of water at 70°C?

b. Will 400 g of sucrose dissolve in a teapot that contains 200 g of water at 70°C? Explain.

Report Sheet - Lab 10

C. Concentration of a Sodium Chloride Solution

C.1 Mass of evaporating dish _____ g

C.2 Volume of NaCl solution _____ mL

C.3 Mass of dish and NaCl solution _____ g

C.4 Mass of dish and dry NaCl _____ g

Calculations

C.5 Mass of NaCl solution _____ g

C.6 Mass of the dry NaCl salt _____ g

C.7 Mass/mass percent _____ % (m/m)

(Show calculations.)

C.8 Mass/volume percent _____ % (m/v)

(Show calculations.)

C.9 Moles of NaCl _____ moles

(Show calculations.)

C.10 Volume of sample in liters _____ L

Laboratory 10

Report Sheet - Lab 10

C.11 Molarity of NaCl solution _____ M

(Show calculations.)

Questions and Problems

Q.3 A 15.0-mL sample of NaCl solution has a mass of 15.78 g. After the NaCl solution is evaporated to dryness, the dry salt residue has a mass of 3.26 g. Calculate the following concentrations for the NaCl solution.

a. % (m/m)

b. % (m/v)

c. molarity (M)

Q.4 How many grams of KI are in 25.0 mL of a 3.0 % (m/v) KI solution?

Q.5 How many milliliters of a 2.5 M $MgCl_2$ solution contain 17.5 g $MgCl_2$?

11

Testing for Cations and Anions

Goals

- Determine the presence of a cation or anion by a chemical reaction.
- Determine the presence of some cations and anions in an unknown salt.

Text References

Review the following before performing this laboratory.

Timberlake; *General, Organic, and Biological Chemistry: Structures of Life*, Third Edition
Sections 5.1 to 5.4; Pages 169 to 183

or

Timberlake; *Chemistry: An Introduction to General, Organic, and Biological Chemistry*, Tenth Edition
Sections 4.1 to 4.4; Pages 121 to 138

Discussion

Solutions such as milk, coffee, tea, and orange juice contain an assortment of ions. In chemical reactions, these ions can give a distinctive flame test, undergo color changes, or form a gas or an insoluble solid. Your observations of the results of a test are the key to identifying those same ions when you test unknown solutions. After you look at a test that produces some change when a particular ion reacts, you will look for the same change in an unknown sample. If the test result is the same, you assume that the ion is present in the unknown. If the same test result does not occur then the ion is not in the unknown: the test is negative.

A. Tests for Positive Ions (Cations)

In this experiment, you will observe the tests for the following cations contained in the indicated solutions:

Cations	Solution
sodium (Na^+)	0.1 M NaCl
potassium (K^+)	0.1 M KCl
calcium (Ca^{2+})	0.1 M $CaCl_2$
iron (Fe^{3+})	0.1 M $FeCl_3$
ammonium (NH_4^+)	0.1 M NH_4Cl

The presence of Na^+, K^+, and Ca^{2+} is determined by the distinctive color the ions give in flame tests. The calcium ion also reacts with ammonium oxalate, $(NH_4)_2C_2O_4$, to give a white precipitate. When NH_4^+ is converted to ammonia (NH_3), a distinctive odor is emitted, and the vapor turns red litmus paper to blue. Iron (Fe^{3+}) is detected by the distinctive red color it gives with potassium thiocyanate, KSCN.

Laboratory 11

B. Tests for Negative Ions (Anions)

In this experiment, you will observe the tests for the following anions contained in the indicated solutions:

Anions	Solution
chloride (Cl^-)	0.1 M NaCl
phosphate (PO_4^{3-})	0.1 M Na_3PO_4
sulfate (SO_4^{2-})	0.1 M Na_2SO_4
carbonate (CO_3^{2-})	0.1 M Na_2CO_3

When $AgNO_3$ is added to the solutions to be tested for anions, several insoluble salts form. However, when nitric acid, HNO_3, is added to these precipitates, all the solids except AgCl will dissolve; the insoluble salt AgCl will remain in the test tube. Phosphate ion (PO_4^{3-}) reacts with ammonium molybdate and forms a yellow solid. When barium chloride, $BaCl_2$, is added to a solution containing SO_4^{2-}, a precipitate of $BaSO_4$ forms. Barium may form insoluble salts with some other anions, but the addition of HNO_3 dissolves all barium salts except $BaSO_4$. The insoluble $BaSO_4$ remains in the test tube after HNO_3 is added. Carbonate anion (CO_3^{2-}) is identified by adding HCl, which produces bubbles of CO_2 gas.

C. Writing the Formula of Your Unknown Salt

As you proceed, you will test known solutions that contain a particular ion. Each student will receive an unknown sample containing a cation and an anion. After you observe reactions of knowns, you will carry out the same tests with your unknown. Therefore, you should expect to see a test that matches the reactions of one of the cations, and another for one of the anions. By identifying the ions in your unknown, you will be able to write the name and formula of your unknown salt.

Lab Information

Time: $2^1/_2$ hr

Comments: Use several small beakers to hold the reagents. Be sure to label each.
HCl and HNO_3 are strong acids, and NaOH is a strong base. Handle with care!
If they are spilled on the skin, rinse thoroughly with water for 10 minutes.
Dispose of test resultant solutions and solids as indicated by the instructor .
Tear out the report sheets and place them beside the matching procedures.

Related Topics: Ions, chemical change, solubility rules

Experimental Procedures *GOGGLES REQUIRED!*

A. Tests for Positive Ions (Cations)

Testing for Cations and Anions

Materials: Spot plate, Bunsen burner and/or hot plate, test tubes and test tube rack, flame-test wire, 4 small beakers, 3 M HCl, 6 M HNO$_3$, 6 M NaOH, dropper bottles of 0.1 M NaCl, 0.1 M KCl, 0.1 M CaCl$_2$, 0.1 M (NH$_4$)$_2$C$_2$O$_4$ (ammonium oxalate), 0.1 M NH$_4$Cl, 0.1 M FeCl$_3$, 0.1 M KSCN (potassium thiocyanate), red litmus paper, warm water bath, stirring rod

Section	Step	Procedure	Check when finished
A	1	In many parts of this experiment, you will be using 2 mL of different solutions. Place 2 mL of water in the same size test tube that you will be using. As you obtain solutions for the experiment, take a volume that matches the height of the 2 mL of water. Be sure to label each test tube. Many test tubes have a frosted section to write on. If not, use a marking pencil or a label.	
A	2	Place about 15 mL of 3 M HCl in a small beaker and label it.	
A	3	Place about 15 mL of 6 M HNO$_3$ in a small beaker and label it.	
A	4	Place about 15 mL of 6 M NaOH in a small beaker and label it.	
A	5	Have another beaker of distilled water for rinsing out the droppers.	
A	6	Preparation of Unknowns: Take a small beaker to your instructor for a sample of the unknown salt solution assigned to you.	
A	7	Record the sample number.	
A	8	If the unknown is a solid, dissolve 1 g of the salt in 50 mL of distilled water. Use small portions of this *unknown* solution in each of the tests. You will need to identify the presence of a cation and an anion in the unknown by comparing the test results of the unknown with the test results given by each of the known solutions.	

A.1 **Flame tests for Na$^+$, K$^+$, and Ca^{2+}**

Section	Step	Procedure	Check when finished
A.1	1	Obtain a spot plate.	
A.1	2	Place 5 to 8 drops of 0.1 M solution of NaCl into a separate well.	
A.1	3	Place 5 to 8 drops of 0.1 M solution of KCl into a separate well.	
A.1	4	Place 5 to 8 drops of 0.1 M solution of CaCl$_2$ into a separate well.	
A.1	5	Place 5 to 8 drops of 0.1 M solution of your unknown into a separate well.	
A.1	6	Dip the test wire in 3 M HCl and heat the wire until the flame is light blue. This will clean the wire.	

Laboratory 11

Section	Step	Procedure	Check when finished
A.1	7	Place the wire loop in the NaCl solution and then into the flame.	
A.1	8	Record the color produced by the Na^+ ion.	
A.1	9	Clean the wire as in step 6.	
A.1	10	Place the wire loop in the KCl solution and then into the flame. The color (pink-lavender) of K^+ does not last long so look for it immediately.	
A.1	11	Record the color produced by the K^+ ion.	
A.1	12	Clean the wire as in step 6.	
A.1	13	Place the wire loop in the $CaCl_2$ solution and then into the flame.	
A.1	14	Record the color produced by the Ca^{2+} ion.	
A.1	15	Testing the unknown: Clean the wire as in step 6.	
A.1	16	Place the wire loop in the unknown solution, then into the flame.	
A.1	17	If there is a color that matches the color of an ion you tested in the flame tests (A.1), you can conclude that you have one of the ions Na^+, K^+, or Ca^{2+} in your unknown. Record.	
A.1	18	If you think Ca^{2+} is present, you may wish to confirm it with test A.2. If the flame test does not produce any color, then Na^+, K^+, and Ca^{2+} are not present in your unknown.	

A.2 Test for calcium ion, Ca^{2+}

Section	Step	Procedure	Check when finished
A.2	1	Place 2 mL of 0.1 M $CaCl_2$ in a test tube.	
A.2	2	Place 2 mL of your unknown solution in another test tube.	
A.2	3	Add 15 drops of ammonium oxalate solution, 0.1 M $(NH_4)_2C_2O_4$, to each.	
A.2	4	Look for a cloudy, white solid (precipitate).	
A.2	5	If the solution remains clear, place the test tube in a warm water bath for 5 minutes, then look for a precipitate. The net equation for the reaction is: $Ca^{2+} + C_2O_4^{2-} \rightarrow CaC_2O_4(s)$ *A white precipitate indicates the presence of Ca^{2+}.*	

A.3 Test for ammonium ion, NH_4^+

Section	Step	Procedure	Check when finished
A.3	1	Place 2 mL of 0.1 M NH_4Cl in a test tube.	
A.3	2	Place 2 mL of your unknown in another test tube.	

Testing for Cations and Anions

Section	Step	Procedure	Check when finished
A.3	3	Add 15 drops of 6 M NaOH to each.	
A.3	4	*Carefully* fan the vapors from the test tube toward you. You may notice the odor of ammonia.	
A.3	5	Place a strip of moistened red litmus paper across the top of the test tube and set the test tube in a warm water bath. Do not let the litmus paper touch the NaOH. The $NH_3(g)$ given off will turn the red litmus paper blue. $$NH_4^+ + OH^- \rightarrow NH_3(g) + H_2O$$ *Ammonia*	
A.3	6	Record results.	

A.4 **Test for ferric ion, Fe^{3+}**

Section	Step	Procedure	Check when finished
A.4	1	Place 2 mL of 0.1 M $FeCl_3$ in a test tube.	
A.4	2	Place 2 mL of your unknown in another test tube.	
A.4	3	Add 5 drops of 6 M HNO_3 to each test tube.	
A.4	4	Add 2–3 drops of potassium thiocyanate, 0.1 M KSCN to each test tube.	
A.4	5	A deep red color indicates that Fe^{3+} is present. A faint pink color is *not* a positive test for iron. Repeat the test with your unknown. Record results. $Fe^{3+} + SCN^- \rightarrow Fe(SCN)^{2+}$ *Deep red color*	
A.4	6	Record results.	

B. Tests for Negative Ions (Anions)

Materials: Test tubes, test tube rack, 0.1 M NaCl, 0.1 M $AgNO_3$ (dropper bottle), 3 M HCl, 6 M HNO_3, stirring rod, 0.1 M Na_2SO_4, 0.1 M $BaCl_2$, 0.1 M Na_3PO_4, $(NH_4)_2MoO_4$ (ammonium molybdate reagent), 0.1 M Na_2CO_3, hot water bath

B.1 **Test for chloride ion, Cl^-**

Section	Step	Procedure	Check when finished
B.1	1	Place 2 mL of 0.1 M NaCl solution in a test tube.	
B.1	2	Place 2 mL of your unknown in another test tube.	
B.1	3	Add 5–10 drops of 0.1 M $AgNO_3$ to each test tube.	
B.1	4	Add 10 drops of 6 M HNO_3 to each test tube.	
B.1	5	Stir with a glass stirring rod. ***Caution: $AgNO_3$ stains the skin.***	

Laboratory 11

		Any white solid that remains is AgCl(s). Any white solids that dissolve with HNO_3 do not contain Cl^-.		
B.1	6	Record the results of your known and unknown. $$Ag^+ + Cl^- \rightarrow AgCl(s)$$ *White precipitate remains after HNO_3 is added.*		

B.2   Test for sulfate ion, SO_4^{2-}

Section	Step	Procedure	Check when finished
B.2	1	Place 2 mL of 0.1 M Na_2SO_4 solution in a test tube.	
B.2	2	Place 2 mL of your unknown in another test tube.	
B.2	3	Add 1 mL (20 drops) of $BaCl_2$ to each test tube.	
B.2	4	Add 5–6 drops of 6 M HNO_3 to each test tube. $BaSO_4$, a white precipitate, does not dissolve in HNO_3. Other anions, CO_3^{2-} and PO_4^{3-}, will also form barium compounds, $Ba_3(PO_4)_2$ and $BaCO_3$, but they will *dissolve* in HNO_3.	
B.2	5	Record results.	

B.3   Test for phosphate ion, PO_4^{3-}

Section	Step	Procedure	Check when finished
B.3	1	Place 2 mL of 0.1 M Na_3PO_4 solution in a test tube.	
B.3	2	Place 2 mL of your unknown in another test tube.	
B.3	3	Add 10 drops of 6 M HNO_3 to each.	
B.3	4	Warm the test tubes in a hot water bath (60°C).	
B.3	5	Add 15 drops of ammonium molybdate solution, $(NH_4)_2MoO_4$. The formation of a yellow precipitate indicates the presence of PO_4^{3-}.	
B.3	6	Record the test results of the known and the unknown.	

B.4   Test for carbonate ion, CO_3^{2-}

Section	Step	Procedure	Check when finished
B.4	1	Place 2 mL of 0.1 M Na_2CO_3 solution in a test tube.	

Testing for Cations and Anions

Section	Step	Procedure	Check when finished	
B.4	2	Place 2 mL of your unknown in another test tube.		
B.4	3	While carefully observing the solution, add 10 drops of 3 M HCl to each test tube. Watch for a strong evolution of bubbles of CO_2 gas as you add the HCl. The gas bubbles are formed quickly, and may be overlooked.		
B.4	4	If gas bubbles were not observed, add another 15–20 drops of HCl *as you watch the solution*.		
B.4	5	Record your results for the known and the unknown. $$Na_2CO_3(aq) + 2HCl(aq) \rightarrow CO_2(g) + H_2O + 2NaCl(aq)$$ Gas bubbles		

C. Writing the Formula of Your Unknown Salt

Your unknown solution was made from a salt composed of a cation and an anion. From your test results, you can identify one of the cations (Na^+, K^+, Ca^{2+}, NH_4^+, or Fe^{3+}) and one of the anions (Cl^-, SO_4^{2-}, PO_4^{3-}, or CO_3^{2-}). For example, if you found that in the cation tests you got the same test result as for Ca^{2+} and in the anion tests you got the same result as for Cl^-, then the ions in your unknown salt would be Ca^{2+} and Cl^-. The formula $CaCl_2$ is written using charge balance.

Section	Step	Procedure	Check when finished
C.1	1	Write the symbols and names of the cation and anion that were present in your unknown.	
C.2	1	Use the ionic charges of the cation and anion to write the formula and name of the salt that was your unknown.	

Laboratory 11

Testing for Cations and Anions

Report Sheet - Lab 11

Date _____   Name _____
Section _____   Team _____
Instructor _____   _____

Pre-Lab Study Questions

1. How can the presence of an ion in a solution be detected?

2. If a reaction produces an insoluble salt, what will you notice happening in the test tube?

3. What tests would you use to confirm that you have a solution of Ag_2PO_4?

4. What tests would you use to confirm that you have a solution of $FeCl_3$?

Laboratory 11

Testing for Cations and Anions

Report Sheet - Lab 11

Date _____ Name _____
Section _____ Team _____
Instructor _____   _____

A. Tests for Positive Ions (Cations)

Unknown number _____

Procedure	Cation Tested	Observations for known	Observations for Unknown
A.1 **Flame tests**	Na^+		
	K^+		
	Ca^{2+}		
A.2 **Oxalate test**	Ca^{2+}		
A.3 **Ammonium test**	NH_4^+		
A.4 **Iron test**	Fe^{3+}		

Identification of the positive ion in the unknown solution

From your test results, what positive ion (cation) is present in your unknown? _____
Explain your choice.

Laboratory 11

Report Sheet - Lab 11

B. Tests for Negative Ions (Anions)

Unknown number _____

Procedure	Anion Tested	Observations for known	Observations for Unknown
B.1 **Chloride test**	Cl^-		
B.2 **Sulfate test**	SO_4^{2-}		
B.3 **Phosphate test**	PO_4^{3-}		
B.4 **Carbonate test**	CO_3^{2-}		

Identification of the negative ion in the unknown solution

From your test results, what negative ion (anion) is present in your unknown? _____

Explain your choice.

C. Writing the Formula of Your Unknown Salt

Unknown sample number _____

C.1   Cation _____   Name _____

       Anion _____   Name _____

C.2   Formula of the unknown salt _____

       Name of the unknown salt _____

12
Solutions, Colloids, and Suspensions

Goals
- Perform chemical tests for chloride, glucose, and starch.
- Use dialysis to distinguish between solutions and colloids.
- Separate colloids from suspensions.
- Discuss the effects of hypotonic and hypertonic solutions on red blood cells.

Text References

Review the following before performing this laboratory.

Timberlake; *General, Organic, and Biological Chemistry: Structures of Life*, Third Edition
Section 8.6; Pages 322 to 328

or

Timberlake; *Chemistry: An Introduction to General, Organic, and Biological Chemistry*, Tenth Edition
Sections 7.6 to 7.7; Pages 265 to 272

Discussion

A. Identification Tests

A chemical test helps us identify the presence or absence of a substance in a solution. In a *positive test*, the chemical change indicates the substance is present. In a *negative test*, there is no chemical reaction and the substance is absent.

B. Osmosis and Dialysis

Osmosis occurs when water moves through semipermeable membranes such as the walls of red blood cells. The osmotic pressure of an *isotonic* solution is the same as that of red blood cells. Both 0.9% NaCl (saline) and 5% glucose solutions are considered isotonic to the cells of the body. In isotonic solution, the flow of water into and out of the red blood cells is equal. A *hypotonic* solution has a lower osmotic pressure than an isotonic solution; the osmotic pressure of a *hypertonic* solution is greater. In either case, the flow of water in and out of the cell is no longer equal, and the cell volume is altered.

When red blood cells are placed in a concentrated salt solution, they form small clumps. This process, called *crenation*, occurs because water diffuses out of the cells into the more concentrated salt solution. If red blood cells are placed in water, they expand and may rupture. This process, called *hemolysis*, occurs because

Laboratory 12

water diffuses into the cells where there is a higher solute concentration. In both cases, osmosis has occurred as water passed through a semipermeable membrane into the more concentrated solution.

In *dialysis*, small particles and water, but not colloids, move across a semipermeable membrane from a high concentration to lower one. Many of the membranes in the body are dialyzing membranes. For example, the intestinal tract consists of a semipermeable membrane that allows the solution particles from digestion to pass into the blood and lymph. Larger, incompletely digested food particles that are colloidal size or larger remain within the intestinal wall. Dialyzing membranes are also used in hemodialysis to separate waste particles, particularly urea, from the blood.

C.   Filtration

In the process of *filtration*, gravity separates suspension particles from the solvent. The pores in the filter paper are smaller than the size of the suspended particles, causing these suspension particles to be trapped on the filter paper. The colloidal particles and the solution particles are smaller and can pass through the pores of the filter paper.

In this experiment, a mixture of starch, NaCl, glucose, and charcoal is poured onto filter paper. Using identification tests, the substances that remain in the filter paper and the substances that pass through the filter paper can be determined.

Lab Information

Time: $2\frac{1}{2}$ hr

Comments: Label the containers to keep track of different solutions.

Tear out the report sheets and place them beside the matching procedures.

Related Topics: Solutions, colloids, suspensions, osmosis, hypertonic solutions, isotonic solutions, hypotonic solutions, dialysis

Dispose of all chemicals safely as directed by your lab instructor.

Experimental Procedures   *GOGGLES REQUIRED!*

A.   Identification Tests

Materials: Three small beakers (100–150 mL), test tubes (6), test tube rack, stirring rod, 50-mL graduated cylinder, test tube holder, droppers, boiling water bath using a hot plate, 1% starch, 10% NaCl, 0.1 M $AgNO_3$, 10% glucose, Benedict's reagent, iodine reagent. *Dropper bottles with these reagents may be available in the laboratory.*

In this experiment, you will perform the identification tests for Cl^-, glucose, and starch. To determine the presence or absence of Cl^-, glucose, or starch in later experiments, refer to the results of these tests. For each

Solutions, Colloids, and Suspensions

test, observe and record the initial properties of the reagent and the final appearance of the solution after the reagent is added. In a positive test, there will be a change in the original properties of the reagent, such as a color change and/or the formation of a precipitate (opaque solid). If there is no change in the appearance of the reagent, the test is *negative*.

Section	Step	Procedure	Check when finished
A	1	Place small amounts of the reagents for these identification tests in small beakers or vials for use throughout this experiment. Label the beakers. Keep the reagents at your desk for the remainder of the experiment.	
A	2	Place 3 mL of water in a test tube for volume comparison.	
A	3	Fill a container with 3–4 mL of 0.1 M AgNO$_3$. **Caution: AgNO$_3$ and iodine reagent stain and iodine is toxic!**	
A	4	Fill a container with 20–25 mL of Benedict's reagent.	
A	5	Fill a container with 4–5 mL of iodine reagent.	
A	6	Fill a container with 10 mL of distilled water.	

A.1 **Chloride (Cl$^-$) test**

Section	Step	Procedure	Check when finished
A.1	1	Place 3 mL of 10% NaCl in a test tube.	
A.1	2	Place 3 mL of distilled water in another test tube. The test tube with the water will be the control or comparison sample.	
A.1	3	Add 2 drops of 0.1 M AgNO$_3$ to each and mix by shaking.	
A.1	4	Record your observations.	
A.1	5	Compare the results for the NaCl solution with the results for the water sample.	

A.2 **Starch test**

Section	Step	Procedure	Check when finished
A.2	1	Place 3 mL of 1% starch solution in a test tube.	
A.2	2	Place 3 mL of distilled water in another test tube.	
A.2	3	Add 2–3 drops of iodine reagent to each and mix by shaking.	
A.2	4	Record your observations.	
A.2	5	Compare the results for the starch sample with the results for the water sample.	

Laboratory 12

A.3 Glucose test

Section	Step	Procedure	Check when finished
A.3	1	Place 3 mL of 10% glucose solution in a test tube.	
A.3	2	Place 3 mL of distilled water in another test tube.	
A.3	3	Add 3 mL of Benedict's reagent to the glucose and to the distilled water and mix by shaking.	
A.3	4	Heat both test tubes in a boiling water bath for 5 minutes.	
A.3	5	Record your observations.	
A.3	6	Compare the results for the glucose with the results for the water sample. *Note: Each time you carry out the glucose test, the test tubes containing Benedict's reagent must be heated in a boiling water bath for 5 minutes.*	

B. Osmosis and Dialysis

Materials: Cellophane tube (15–20 cm), test tubes (3), test tube rack, test tube holder, stirring rod, droppers, 50-mL graduated cylinder, funnel, 100-mL beaker, 250-mL beaker, boiling water bath, 10% NaCl, 10% glucose, 1% starch, 0.1 M $AgNO_3$, Benedict's solution, iodine reagent

Section	Step	Procedure	Check when finished
B	1	In a small beaker, combine 10 mL of 10% NaCl, 10 mL of 10% glucose, and 10 mL of 1% starch solution.	
B	2	Tie a knot in one end of a piece of cellophane tubing (dialysis bag).	
B	3	Place a funnel in the open end and pour in about 20 mL of the mixture prepared in B.1. Save the rest of the mixture for part C.	
B	4	Tie a firm knot in the open end to close the dialysis bag.	
B	5	Rinse the outside of the dialysis bag with distilled water.	
B	6	Place the dialysis bag in a 250-mL beaker and cover with distilled water. See Figure 12.1.	

Solutions, Colloids, and Suspensions

Figure 12.1 Dialysis bag placed in distilled water

Zero minute test

Section	Step	Procedure	Check when finished
B.1	1	Pour off 15 mL of the solution from the outside of the dialysis bag into a small beaker.	
B.1	2	Divide the 15 mL into three test tubes (5 mL each).	
B.1	3	Repeat A.1 step 3 to step 5 on the first tube.	
B.1	4	Record the result as positive if chloride is present and negative if it is not.	
B.1	5	Repeat A.2 step 3 to step 5 on the second tube.	
B.1	6	Record the result as positive if starch is present and negative if it is not.	
B.1	7	Repeat A.3 step 3 to step 6 on the third tube.	
B.1	8	Record the result as positive if glucose is present and negative if it is not.	

20 minute tests

Section	Step	Procedure	Check when finished
B.2	1	Pour off 15 mL of the solution from the outside of the dialysis bag into a small beaker.	
B.2	2	Divide the 15 mL into three test tubes (5 mL each).	
B.2	3	Repeat A.1 step 3 to step 5 on the first tube.	
B.2	4	Record the result as positive if chloride is present and negative if it is not.	
B.2	5	Repeat A.2 step 3 to step 5 on the second tube.	

Laboratory 12

Section	Step	Procedure	Check when finished
B.2	6	Record the result as positive if starch is present and negative if it is not.	
B.2	7	Repeat A.3 step 3 to step 6 on the third tube.	
B.2	8	Record the result as positive if glucose is present and negative if it is not.	

40 minute test

Section	Step	Procedure	Check when finished
B.3	1	Pour off 15 mL of the solution from the outside of the dialysis bag into a small beaker.	
B.3	2	Divide the 15 mL into three test tubes (5 mL each).	
B.3	3	Repeat A.1 step 3 to step 5.	
B.3	4	Record the result as positive if chloride is present and negative if it is not.	
B.3	5	Repeat A.2 step 3 to step 5.	
B.3	6	Record the result as positive if starch is present and negative if it is not.	
B.3	7	Repeat A.3 step 3 to step 6.	
B.3	8	Record the result as positive if glucose is present and negative if it is not.	

40 minute test – contents of bag

Section	Step	Procedure	Check when finished
B.4	1	Open the dialysis bag and pour off 15 mL of the solution from the inside of the dialysis bag into a small beaker. *Save the remainder of the bag's contents for use in part C.*	
B.4	2	Divide the 15 mL into three test tubes (5 mL each).	
B.4	3	Repeat A.1 step 3 to step 5 on the first tube.	
B.4	4	Record the result as positive if chloride is present and negative if it is not.	
B.4	5	Repeat A.2 step 3 to step 5 on the second tube.	
B.4	6	Record the result as positive if starch is present and negative if it is not.	
B.4	7	Repeat A.3 step 3 to step 6 on the third tube.	
B.4	8	Record the result as positive if glucose is present and negative if it is not.	

Solutions, Colloids, and Suspensions

C. Filtration

Materials: Funnel, filter paper, 50-mL graduated cylinder, test tubes (3), test tube rack, test tube holder, stirring rod, droppers, two 150-mL beakers, boiling water bath, powdered charcoal, iodine reagent, 0.1 M $AgNO_3$, Benedict's solution.

To the remaining mixture of glucose, NaCl, and starch from part B.3, add a small amount of powdered charcoal and mix.

Section	Step	Procedure	Check when finished
C	1	Fold a piece of filter paper in half and in half again.	
C	2	Open the fold to give a cone-like shape and place the filter paper cone in a funnel.	
C	3	Push it gently against the sides and moisten with water.	
C	4	Place a small beaker under the funnel.	
C	5	Pour the mixture into the filter paper.	
C	6	Collect the liquid (filtrate) that passes through the filter. See Figure 12.2.	

Figure 12.2 Filtering a mixture

C.1 Describe what you see in the filter paper.

C.2 Filtration

Section	Step	Procedure	Check when finished
C.2	1	Divide 15 mL of the material in the beaker under the filter into three test tubes (5 mL each).	
C.2	2	Repeat A.1 step 3 to step 5 on the first tube.	

161

Laboratory 12

Section	Step	Procedure	Check when finished
C.2	3	Record the result as positive if chloride is present and negative if it is not.	
C.2	4	Repeat A.2 step 3 to step 5 on the second tube.	
C.2	5	Record the result as positive if starch is present and negative if it is not.	
C.2	6	Repeat A.3 step 3 to step 6 on the third tube.	
C.2	7	Record the result as positive if glucose is present and negative if it is not.	

Section	Step	Procedure	Check when finished
C.3	1	From the test results, determine what was trapped in the filter and what passed through. Identify the substance that is a suspension.	

Report Sheet - Lab 12

Date _____ Name _____

Section _____ Team _____

Instructor _____ _____

Pre-Lab Study Questions

1. In making pickles, a cucumber is placed in a concentrated salt solution. Explain what happens.

2. Why is it important that cell membranes are semipermeable membranes?

3. What is the difference between osmosis and dialysis?

4. How does an artificial kidney separate waste products from the blood?

5. In the experimental procedure B, predict which substances, glucose, starch or NaCl will pass through the filter paper and which will stay on the filter paper.

6. In the experimental procedure C, predict which substances, glucose, starch or NaCl will pass through the dialysis bag and into the solution.

Report Sheet - Lab 12

Date _____   Name _____
Section _____   Team _____
Instructor _____   _____

A. Identification Tests

Test	Reagent Added	Results of Positive Test	Results with Water Control
A.1 Cl⁻	AgNO$_3$		
A.2 Starch	Iodine		
A.3 Glucose	Benedict's; heat		

B. Osmosis and Dialysis

Test Results of Water Outside Dialysis Bag

Time	Cl⁻ Present?	Starch Present?	Glucose Present?
0 minutes			
20 minutes			
40 minutes			
Contents of dialysis bag			

Questions and Problems

Q.1 Which substance(s) were found in the water *outside* the dialysis bag?

Q.2 How did those substance(s) go into the water outside the dialysis bag?

Report Sheet - Lab 12

Q.3 What substance(s) were retained inside the dialysis bag? Why were they retained?

C. Filtration

C.1 Appearance of filter paper _____

   Substance present   _____

C.2 Test for	Results of Test	Substance Present in Filtrate?
Cl^-		
Starch		
Glucose		

C.3 Which substance is a suspension? _____

   Which substances are solutions or colloids? _____

Questions and Problems

Q.4 What is an isotonic solution?

   What is a hypotonic solution?

   What is a hypertonic solution?

Solutions, Colloids, and Suspensions

Report Sheet - Lab 12

Q.5   State whether each of the following are isotonic, hypotonic, or hypertonic:

a. H_2O _____

b. 0.9% NaCl _____

c. 10% glucose _____

d. 3% NaCl _____

e. 0.2% NaCl _____

Q.6   A red blood cell in a hypertonic solution will shrink in volume as it undergoes *crenation*. In a hypotonic solution, a red blood cell will swell and possibly burst as it undergoes *hemolysis*. Predict the effect on a red blood cell (crenation, hemolysis, or none) that the following solutions would have:

a.   2% NaCl _____

b. H_2O _____

c.   5% glucose _____

d.   1% glucose _____

e.   10% glucose _____

Q.7   A parenteral solution is a solution that is injected into the tissues or bloodstream, but not given orally. Why are isotonic solutions used as parenteral solutions?

13
Acids and Bases

Goals

- Prepare a naturally occurring dye from red cabbage to use as a pH indicator.
- Measure the pH of several substances using cabbage indicator and a pH meter.
- Calculate pH from the [H$^+$] or the [OH$^-$] of a solution.
- Calculate the molar concentration and percentage of acetic acid in vinegar.
- Calculate pH from the [H$^+$] or the [OH$^-$] of a solution.
- Prepare a sample for titration with a base.
- Set up a buret and perform a titration using appropriate technique.
- Calculate the molar concentration and percentage of acetic acid in vinegar.

Text References

Review the following before performing this laboratory.

Timberlake; *General, Organic, and Biological Chemistry: Structures of Life*, Third Edition
Sections 10.1 to 10.5; Pages 372 to 398

or

Timberlake; *Chemistry: An Introduction to General, Organic, and Biological Chemistry*, Tenth Edition
Section 8.5 to 8.3; Pages 298 to 306

Discussion

An acid is a substance that dissolves in water and donates a hydrogen ion, or proton (H$^+$), to water. In the laboratory we have been using acids such as hydrochloric acid (HCl) and nitric acid (HNO$_3$).

$$HCl + H_2O \rightarrow H_3O^+ + Cl^-$$
hydronium ion

You use acids and bases every day. There are acids in oranges, lemons, vinegar, and bleach. In this experiment we will use acetic acid (CH$_3$CO$_2$H). Acetic acid is the acid in vinegar that gives it a sour taste.

A base is a substance that accepts a proton. Some household bases include ammonia, detergents, and oven-cleaning products. Some typical bases used in the laboratory are sodium hydroxide (NaOH) and potassium hydroxide (KOH). Most of the common bases dissolve in water and produce hydroxide ions, OH$^-$.

$$NaOH \rightarrow Na^+ + OH^-$$

Laboratory 13

An important weak base found in the laboratory and in some household cleaners is ammonia. In water, ammonia reacts to form ammonium and hydroxide ions:

$$NH_3 + H_2O \rightarrow NH_4^+ + OH^-$$

In a neutralization reaction, the protons (H^+) from the acid combine with hydroxide ions (OH^-) from the base to produce water (H_2O). The remaining substance is a salt, which is composed of ions from the acid and base. For example, the neutralization of HCl by NaOH is written as

$$HCl + NaOH \rightarrow NaCl + H_2O$$

If we write the ionic substances in the equation as ions, we see that the H^+ and the OH^- form water.

$$H^+ + Cl^- + Na^+ + OH^- \rightarrow Na^+ + Cl^- + H_2O$$
$$H^+ + OH^- \rightarrow H_2O$$

In a complete neutralization, the amount of H^+ will be equal to the amount of OH^-.

A. pH Color Using Red Cabbage Indicator

The pH of a solution tells us whether a solution is acidic, basic, or neutral. On the pH scale, pH values below 7 are acidic, equal to 7 is neutral, and values above 7 are basic. Typically, the pH scale has values between 0 and 14.

pH scale

0 1 2 3 4 5 6 7 8 9 10 11 12 13 14

←———— acidic ————→ neutral ←———— basic ————→

Many natural substances contain dyes that produce distinctive colors at different pH values. By extracting (removing) the dye from red cabbage leaves, a natural indicator can be prepared. Adding the red cabbage solution to solutions of a variety of acids and bases will produce a series of distinctive colors. When the red cabbage solution is added to a test sample, the color produced can be matched to the colors of the pH reference set to determine the pH of the sample. A pH meter can also be used to measure pH.

B. Measuring pH

The concentration (moles/liter, indicated by brackets [ ]) of H_3O^+ or OH^- can be determined from the ionization constant for water (K_w). In pure water, $[H_3O^+] = [OH^-] = 1 \times 10^{-7}$ M at 25°C.

$$K_w = [H_3O^+][OH^-] = [1 \times 10^{-7}][1 \times 10^{-7}] = 1 \times 10^{-14}$$

If the $[H_3O^+]$ or $[OH^-]$ for an acid or a base is known, the other can be calculated. For example, an acid has a $[H^+] = 1 \times 10^{-4}$ M. We can find the $[OH^-]$ of the solution by solving the K_w expression for $[OH^-]$:

$$[OH^-] = \frac{1 \times 10^{-14}}{[H_3O^+]} = \frac{1 \times 10^{-14}}{1 \times 10^{-4}} = 1 \times 10^{-10} \text{ M}$$

The pH of a solution is a measure of its $[H_3O^+]$. It is defined as the negative log of the hydrogen ion concentration.

$$pH = -\log [H_3O^+]$$

Therefore, a solution with a $[H_3O^+] = 1 \times 10^{-4}$ M has a pH of 4.0, and is acidic.
A solution with a $[H_3O^+] = 1 \times 10^{-11}$ M has a pH of 11.0, and is basic.

C. Acetic Acid in Vinegar

Vinegar is an aqueous solution of acetic acid, HC2H3O2 or CH3COOH. The amount of acetic acid in a vinegar solution can be determined by neutralizing the acid with a base, in this case NaOH. As shown in the following equation, one mole of acetic acid is neutralized by one mole of NaOH.

$$CH_3CO_2H + NaOH \rightarrow CH_3CO_2^- \, Na^+ + H_2O$$
Acetic acid   Base   Salt

A titration involves the addition of a specific amount of base required to neutralize an acid in a sample. When all the H^+ (or H_3O^+) from the acid has been neutralized, an indicator in the sample will change color. This change in the indicator color determines the endpoint, which signals that the addition of the base should be stopped. The volume of base used to neutralize the acid is then determined. In this experiment, phenolphthalein is the indicator; it changes from colorless in acid to a faint but permanent pink color in base.

Calculating the molarity of acetic acid

Using the average measured volume of the NaOH, and its molarity (on the label), the moles of NaOH used can be calculated.

$$\text{Moles NaOH used} = \text{L NaOH used} \times \frac{\text{moles NaOH}}{1 \text{ L NaOH}}$$

When an acid is completely neutralized, the moles of NaOH are equal to the moles of acetic acid (HC2H3O2) present in the sample. This occurs because there are the same number of H+ and OH– ions in the reactants:

$$\text{Moles } HC_2H_3O_2 = \text{moles NaOH}$$

Using the moles of acid, the molarity of acetic acid in the 5.0-mL sample of vinegar is calculated.

$$\text{Molarity (M) } HC_2H_3O_2 = \frac{\text{moles } HC_2H_3O_2}{0.0050 \text{ L vinegar}}$$

Calculating the percent (mass/volume) of acetic acid

To calculate the percent (m/v) of $HC_2H_3O_2$ in vinegar, we convert the moles of acetic acid to grams using the molar mass of acetic acid, 60.0 g/mole.

Laboratory 13

$$g\ HC_2H_3O_2 = moles\ HC_2H_3O_2 \times \frac{600\ g\ HC_2H_3O_2}{1\ mole\ HC_2H_3O_2}$$

$$Percent\ (m/v) = \frac{g\ HC_2H_3O_2}{5.0\ mL} \times 100$$

Lab Information

Time: 3 hr

Comments: Students may be asked to bring a red cabbage to class.

Share test tubes with your lab neighbors to prepare the pH reference solutions.

Tear out the report sheets and place them beside the matching procedures.

Related Topics: Acids, bases, pH, titration

Dispose of all chemicals as directed by your lab instructor.

Experimental Procedures GOGGLES REQUIRED!

A. pH Color Using Red Cabbage Indicator

Materials: Red cabbage leaves, 400-mL beaker, distilled water, Bunsen burner or hot plate or blender or food processor, 150-mL beaker, test tubes, two test tube racks, set of buffers with pH ranging from 1 to 13

A.1 **Preparing red cabbage indicator- 1.** There are two alternative procedures to produce a pH indicator from red cabbage. The first procedure involves boiling the cabbage leaves in water.

Section	Step	Procedure	Check when finished
A.1	1	Place 5 or 6 torn leaves from red cabbage in a 400-mL beaker.	
A.1	2	Add about 150–200 mL of distilled water to cover the leaves.	
A.1	3	Heat on a hot plate or using a Bunsen burner, but do not boil.	
A.1	4	When the solution has attained a dark purple color, turn off the burner and cool.	
A.1	5	Retain about 100 mL of the indicator (red cabbage) solution.	

Acids and Bases

A.2 **Preparing red cabbage indicator- 2.** The second procedure involves blending water with cabbage leaves in a high-speed kitchen blender.

Section	Step	Procedure	Check when finished
A.2	1	Place 5 or 6 torn leaves from red cabbage in a blender.	
A.2	2	Add about 150–200 mL of distilled water to cover the leaves.	
A.2	3	Blend the leaves and water together at high speed.	
A.2	4	Filter the resultant solution.	
A.2	5	Retain about 100 mL of the indicator (red cabbage) solution.	

A.3 **Preparing pH reference standards**

Section	Step	Procedure	Check when finished
A.3	1	Using a 150-mL beaker, obtain 50 mL of cabbage dye indicator.	
A.3	2	Arrange 13 test tubes in two test tube racks. You may need to combine your test tube set with your neighbor's set. (Your instructor may prepare a pH reference set for the entire class.)	
A.3	3	Pour 3–4 mL of each buffer in a separate test tube to create a set with pH values of 1–13. Caution: Low pH values are strongly acidic; high pH values are strongly basic. Work with care.	
A.3	4	To each test tube, add 2–3 mL of the cooled red cabbage solution. If you wish a deeper color, add more cabbage solution. Shake test tube to mix.	
A.3	5	Describe the colors of the pH solutions.	
A.3	6	Keep this reference set for the next part of the experiment.	

B. Measuring pH

Materials: Shell vials or test tubes, samples to test for pH (shampoo, conditioner, mouthwash, antacids, detergents, fruit juice, vinegar, cleaners, aspirin, etc.), cabbage juice indicator from part A, pH meter, calibration buffers, wash bottle, Kimwipes™

B.1 **Using red cabbage indicator to measure pH**

Section	Step	Procedure	Check when finished
B.1	1	Place 3–4 mL of a sample in a shell vial (or a test tube).	
B.1	2	Add 2–3 mL of red cabbage solution.	
B.1	3	Describe the color and compare to the colors of the pH reference set. The pH of the buffer in the reference set that gives the best color match is the pH of the sample.	

Laboratory 13

B.1	4	Record.					
B.1	5	Repeat steps 1 to 4 for various samples					

Using a pH meter to measure pH. Your instructor will demonstrate the use of the pH meter and calibrate it with a known pH buffer

Section	Step	Procedure	Check when finished			
B.2	1	After you determine the pH of a sample using the red cabbage solution, take the sample to a pH meter, and record the pH.				
B.2	2	Rinse off the electrode with distilled water.				
B.2	3	Repeat steps 1 and 2 for various samples.				

C. Acetic Acid in Vinegar

Materials: Vinegar (white), two beakers (150-and 250-mL), 250-mL Erlenmeyer flask, 10-mL graduated cylinder (or a 5-mL pipet and bulb), phenolphthalein indicator, 50-mL buret (or 25-mL buret), buret clamp, small funnel to fit buret, 0.1 M NaOH (standardized), white paper or paper towel

Section	Step	Procedure	Check when finished			
C.1	1	Obtain about 20 mL of vinegar in a small beaker.				
C.1	2	Record the brand of vinegar and the % acetic acid stated on the label if available.				
C.1	3	Using a 10-mL graduated cylinder or a 5.0-mL pipet, transfer 5.0 mL of vinegar to a 250-mL Erlenmeyer flask. (See the instructions for using a pipet below.)				
C.1	4	Add about 25 mL of distilled water to increase the volume of the solution for titration. This will not affect your results.				
C.1	5	Add 2–3 drops of the phenolphthalein indicator to the solution in the flask.				

Using a pipet: Place the pipet bulb on the end of the pipet and squeeze the bulb to remove air. Place the tip of the pipet in the vinegar in the beaker and allow the bulb to slowly expand. (If the bulb was squeezed too much, the change in pressure will draw liquid up into the bulb.) When the liquid goes above the volume line, but not into the bulb, carefully remove the bulb and quickly place your second finger (index finger) tightly over the end of the stem. By adjusting the pressure of the index finger, lower the liquid to the etched line that marks the 5.0-mL volume and stop. Lift the pipet with its 5.0 mL of vinegar out of the beaker and let it drain into an Erlenmeyer flask. Touch the tip of the pipet to the wall of the flask to remove the rest of the vinegar. A small

Acids and Bases

amount that remains in the tip has been included in the calibration of the pipet. See Figure 13.1 for use of a pipet. Caution: If you are using a pipet, use a suction bulb to draw vinegar into a pipet. Do not pipet by mouth!

Figure 13.1 Using a bulb to draw a liquid into a pipet

Section	Step	Procedure	Check when finished
C.2	1	Obtain a 50-mL or 25-mL buret and place it in a buret clamp or butterfly clamp as shown in Figure 13.2.	
C.2	2	Using a 250-mL beaker, obtain about 100 mL of 0.1 M NaOH solution. (If you are using a 25.0-mL buret, use a 0.2 M NaOH solution.)	
C.2	3	Record the molarity (M) of the NaOH solution that is stated on the label of the reagent bottle.	
C.2	4	Rinse the buret with two 5-mL portions of the NaOH solution.	
C.2	5	Discard the NaOH washings.	

Figure 13.2 Buret setup for acid–base titration

Laboratory 13

Section	Step	Procedure	Check when finished
C.3	1	Observe the markings on the buret. The top is marked 0.0 mL, and 50.0 mL (or 25.0 mL) is marked at the bottom.	
C.3	2	Place a small funnel in the top of the buret and carefully pour NaOH into the funnel.	
C.3	3	Pour slowly as the NaOH fills the buret.	
C.3	4	Lift the funnel and allow the NaOH to go above the top line (0.0 mL).	
C.3	5	Slowly open the stopcock and drain NaOH into a waste beaker until the meniscus is at the 0.00 mL line or below. The buret tip should be full of NaOH solution, and free of bubbles.	
C.3	6	Record the initial buret reading of NaOH.	

Section	Step	Procedure	Check when finished
C.4	1	Place the flask containing the vinegar solution under the buret on a piece of white paper. (Be sure you added indicator. A.1 step 5)	
C.4	2	Begin to add NaOH to the solution by opening and closing the stopcock with your left hand (if you are right-handed).	
C.4	3	Swirl the flask with your right hand to mix the acid and the base. At first, the pink color produced by the reaction will disappear quickly. As you near the endpoint, the pink color will be more persistent and disappear slowly.	
C.4	4	*Slow down* the addition of the NaOH to drops at this time.	
C.4	5	When one drop of NaOH gives a faint, permanent pink color to the sample, *stop adding NaOH*. You have reached the endpoint of the titration. See Figure 13.3.	
C.4	6	At the endpoint, record the final buret reading of the NaOH.	
C.4	7	Repeat A.1 steps 3 to 5, A.3 steps 1 through 6 and A.4 steps 1 through 6 twice more.	

Figure 13.3 During a titration, the solution in the flask is swirled as NaOH is added to the acid sample.

Calculations

Section	Step	Procedure	Check when finished
C.5	1	Calculate the volume of NaOH used to neutralize the vinegar sample(s). Volume (C.5) = Final Volume (NaOH) (C.4) − initial volume (NaOH) (C.3)	
C.5	2	After the titration of two or three samples of vinegar, calculate the average volume of NaOH used. Total the volumes of NaOH used, and divide by the number of samples you used. $$\text{mL (average)} \frac{\text{Volume (1)} + \text{Volume (2)} + \text{Volume (3)}}{3}$$	
C.6	1	Convert the volume (average) of NaOH (C.5) used to a volume in liters (L). L = (C.5)/1000 $$L = \text{Average volume (mL) NaOH} \times \frac{1\ L}{1000\ mL}$$	
C.7	1	Calculate the moles of NaOH using the volume (L) and molarity of the NaOH. Moles NaOH = C.6 × C.2 $$\text{Moles NaOH used} = L\ \text{NaOH used} \times \frac{\text{moles NaOH}}{1\ L\ \text{NaOH}}$$	
C.8	1	Record the moles of acid present in the vinegar, which are equal to the moles of NaOH used. Moles $HC_2H_2O_2$ = A.7 Moles $HC_2H_3O_2$ = moles NaOH used	
C.9	1	Calculate the molarity (M) of the acetic acid ($HC_2H_3O_2$) in the vinegar sample. Molarity (M) = (C.7)/0.005 L $$\text{Molarity (M) } HC_2H_3O_2 = \frac{\text{moles } HC_2H_3O_2}{0.0050\ L\ \text{vinegar}}$$	

Laboratory 13

C.10	1	Using the number of moles of acetic acid, calculate the number of grams of acetic acid in the vinegar sample. The molar mass of acetic acid is 60.0 g/mole. g $HC_2H_2O_2$ = (C.8) × 60 $$g\ HC_2H_3O_2 = moles\ HC_2H_3O_2 \times \frac{60.0\ g\ HC_2H_3O_2}{1\ mole\ HC_2H_3O_2}$$	
C.11	1	Calculate the percent (mass/volume) acetic acid in the vinegar. For an original volume of 5.0 mL of vinegar, the percent is calculated as follows: Percent (m/v) = (C.10) × 100/5 $$Percent\ (m/v) = \frac{g\ HC_2H_3O_2}{5.0\ mL} \times 100$$	

Acids and Bases

Report Sheet - Lab 13

Date _____ Name _____
Section _____ Team _____
Instructor _____   _____

Pre-Lab Study Questions

1. What does the pH of a solution tell you?

2. What is neutralization?

3. Is a solution with a pH of 12 acidic or basic?

4. Is a solution with a pH of 2 acidic or basic?

5. If you add acid or base to water, how do you predict that the pH will change?

Laboratory 13

6. What is neutralization?

7. Write an equation for the neutralization of H2SO4 by KOH.

8. What is a titration?

9. What is the function of an indicator in a titration?

Acids and Bases

Report Sheet - Lab 13

Date _____ Name _____
Section _____ Team _____
Instructor _____   _____

A. pH Colors Using Red Cabbage Indicator

pH	Colors of Acidic Solutions
1	
2	
3	
4	
5	
6	

pH	Color of Neutral Solutions
7	

pH	Colors of Basic Solutions
8	
9	
10	
11	
12	
13	

B. Measuring pH

Substance	Color with Indicator	pH Using Indicator	pH Using pH Meter	Acidic, Basic, or Neutral?
Household cleaners				
vinegar				
ammonia				
Drinks, juices				
lemon juice				
apple juice				

Laboratory 13

Report Sheet - Lab 13

Substance	Color with Indicator	pH Using Indicator	pH Using pH Meter	Acidic, Basic, or Neutral?
Detergents, shampoos				
shampoo				
detergent				
hair conditioner				
Health aids				
mouthwash				
antacid				
aspirin				
Other items				

Questions and Problems

Q.1 Complete the following table:

$[H_3O^+]$	$[OH^-]$	pH	Acidic, Basic, or Neutral?
1×10^{-6}			
		10.0	
	1×10^{-3}		
			Neutral

Q.2 The label on the shampoo claims that it is pH balanced. What do you think "pH balanced" means?

Acids and Bases

Report Sheet - Lab 13

Q.3 A solution has a $[OH^-] = 1 \times 10^{-5}$ M. What are the $[H_3O^+]$ and the pH of the solution?

Q.4 A sample of 0.0020 mole of HCl is dissolved in water to make a 2000-mL solution. Calculate the molarity of the HCl solution, the $[H_3O^+]$, and the pH. For a strong acid such as HCl, the $[H_3O^+]$ is the same as the molarity of the HCl solution.

$$HCl + H_2O \rightarrow H_3O^+ + Cl^-$$

C. Acetic Acid in Vinegar

C.1 Brand _____ Volume <u>5.0 mL</u> (% on label) _____%

C.2 Molarity (M) of NaOH (stated on label) _____ M

		Trial 1	Trial 2	Trial 3
C.3	Initial NaOH level in buret			
C.4	Final NaOH level in buret			
C.5	Volume (mL) of NaOH used			
	Average volume (mL)			
C.6	Average volume in liters (L)			

C.7 Moles of NaOH used in titration _____ mole NaOH
(Show calculations.)

C.8 Moles of $HC_2H_3O_2$ neutralized by NaOH _____ mole $HC_2H_3O_2$

C.9 Molarity of $HC_2H_3O_2$ _____ M $HC_2H_3O_2$
(Show calculations.)

Laboratory 13

C.10 Grams $HC_2H_3O_2$ _____ g $HC_2H_3O_2$

(Show calculations.)

C.11 Percent (m/v) $HC_2H_3O_2$ in vinegar _____ % $HC_2H_3O_2$

(Show calculations.)

Questions and Problems

Q.5 How many grams of $Mg(OH)_2$ will be needed to neutralize 25 mL of stomach acid if stomach acid is 0.10 M HCl?

Q.6 How many mL of a 0.10 M NaOH solution are needed to neutralize 15 mL of 0.20 M H_3PO_4 solution?

14

Alcohols, Aldehydes, and Ketones

Goals

- Determine chemical and physical properties of alcohols and phenols.
- Classify an alcohol as primary, secondary, or tertiary.
- Perform a chemical test to distinguish between the classes of alcohols.
- Write the formulas of the oxidation products of alcohols.
- Write the functional groups of aldehydes and ketones.
- Determine chemical and physical properties of aldehydes and ketones.
- Perform chemical tests to distinguish between aldehydes and ketones.

Text References

Review the following before performing this laboratory.

Timberlake; *General, Organic, and Biological Chemistry: Structures of Life*, Third Edition
Sections 13.1 to 13.4 and 14.1 to 14.4; Pages 480 to 500 and 509 to 524

or

Timberlake; *Chemistry: An Introduction to General, Organic, and Biological Chemistry*, Tenth Edition
Sections 12.1 to 12.5; Pages 406 to 426

Discussion

A. Structures of Alcohols and Phenol

Alcohols are organic compounds that contain the hydroxyl group (–OH). The simplest alcohol is methanol. Ethanol is found in alcoholic beverages and preservatives, and is used as a solvent.

2-Propanol, also known as rubbing alcohol or isopropyl alcohol, is found in astringents and perfumes.

A benzene ring with a hydroxyl group is known as phenol. Concentrated solutions of phenol are caustic and cause burns. However, derivatives of phenol, such as thymol, are used as antiseptics and are sometimes found in cough drops.

Laboratory 14

Phenol

Thymol
(2-isopropyl-5-methylphenol)

Classification of Alcohols

In a primary (1°) alcohol, the carbon atom attached to the –OH group is bonded to one other carbon atom. In a secondary (2°) alcohol, carbon with the –OH is attached to two carbon atoms and in a tertiary (3°) alcohol to three carbon atoms.

Ethanol
primary (1°) alcohol

2-Propanol
secondary (2°) alcohol

2-Methyl-2-Propanol
tertiary (3°) alcohol

B. Properties of Alcohols and Phenol

The polarity of the hydroxyl group (–OH) makes alcohols with four or fewer carbon atoms soluble in water because they can form hydrogen bonds. However, in longer-chain alcohols, a large hydrocarbon portion makes them insoluble in water.

Hydrogen bonding between the hydroxyl group of methanol and water

C. Oxidation of Alcohols

Primary and secondary alcohols are easily oxidized. An oxidation consists of removing an H from the –OH group and another H from the C atom attached to the –OH group. Tertiary alcohols do not undergo oxidation because there are no H atoms on that C atom with the –OH. Primary and secondary alcohols can be distinguished from tertiary alcohols using a solution with chromate, CrO_4^{2-}. An oxidation has occurred when the orange color of the chromate solution turns green.

$$CH_3-CH_2-OH \text{ (1° Alcohol)} + CrO_4^{2-} \text{ (Orange)} \xrightarrow{H^+} CH_3-\underset{\underset{H}{\|}}{\overset{O}{C}}-H \text{ (Aldehyde)} + Cr^{3+} \text{ (Green)}$$

$$CH_3-\underset{\underset{CH_3}{|}}{CH}-OH \text{ (2° Alcohol)} + CrO_4^{2-} \text{ (Orange)} \xrightarrow{H^+} CH_3-\overset{O}{\overset{\|}{C}}-CH_3 \text{ (Ketone)} + Cr^{3+} \text{ (Green)}$$

$$CH_3-\underset{\underset{CH_3}{|}}{\overset{\overset{CH_3}{|}}{C}}-OH \text{ (3° Alcohol)} + CrO_4^{2-} \text{ (Orange)} \xrightarrow{H^+} \text{No reaction (stays orange)}$$

D. Properties of Aldehydes and Ketones

Aldehydes and ketones both contain the carbonyl group. In an aldehyde, the carbonyl group has a hydrogen atom attached; the aldehyde functional group occurs at the end of the carbon chain. In a ketone, the carbonyl group is located between two of the carbon atoms within the chain.

Aldehydes

- Carbonyl functional group
- Acetaldehyde (ethanal): CH_3-CHO
- Propionaldehyde (propanal): CH_3CH_2-CHO

Ketone

- Acetone (2-propanone): $CH_3-CO-CH_3$

Many aldehydes and ketones have sharp odors. If you have taken a biology class, you may have noticed the odor of Formalin™, which is a solution of formaldehyde. When you remove fingernail polish, you may notice the strong odor of acetone, the simplest ketone, which is used as the solvent. Aromatic aldehydes have a variety of odors. Benzaldehyde, the simplest aromatic aldehyde, has an odor of almonds.

Formaldehyde: $H-CHO-H$

Benzaldehyde

E. Iodoform Test for Methyl Ketones

Ketones containing a methyl group attached to the carbonyl give a reaction with iodine (I_2) in a NaOH solution. The reaction produces solid, yellow iodoform, CHI_3. Iodoform, which has a strong medicinal odor, is used as an antiseptic.

Laboratory 14

$$CH_3-\underset{\underset{Methyl\ ketone}{}}{\overset{\overset{O}{\|}}{C}}-CH_3 + \underset{\underset{(red)}{Iodine}}{3I_2} + 4NaOH \longrightarrow CH_3-\overset{\overset{O}{\|}}{C}-O^-Na^+ + \underset{\underset{(yellow)}{Iodoform}}{CHI_3} + 3NaI + 3H_2O$$

F. Oxidation of Aldehydes and Ketones

Aldehydes are oxidized using Benedict's solution, which contains cupric ion, Cu^{2+}. Because ketones cannot oxidize, this test can distinguish aldehydes from most ketones. In the oxidation reaction, the blue-green Cu^{2+} is reduced to cuprous ion (Cu^+), which forms a reddish-orange precipitate of Cu_2O.

$$\underset{Aldehyde}{CH_3-\overset{\overset{O}{\|}}{C}-H} + \underset{Blue}{2Cu^{2+}} \longrightarrow CH_3-\overset{\overset{O}{\|}}{C}-OH + \underset{Red\text{-}orange}{Cu_2O(s)}$$

$$\underset{Ketone}{CH_3-\overset{\overset{O}{\|}}{C}-CH_3} + \underset{Blue}{2Cu^{2+}} \longrightarrow \text{No reaction (stays blue)}$$

Lab Information

Time:	3 hr
Comments:	*Be careful when you work with chromate solution. It contains concentrated acid.*
	Do not use burners in lab when you work with flammable organic compounds.
	Tear out the Lab report sheets and place them beside the matching procedures.
Related topics:	Alcohols, classification of alcohols, solubility of alcohols in water, phenols, oxidation of alcohols

Dispose of all chemicals as directed by your lab instructor.

Alcohols, Aldehydes, and Ketones

Experimental Procedures   *GOGGLES REQUIRED!*

A. Structures of Alcohols and Phenol

Materials: Organic model kits

Section	Step	Procedure	Check when finished
A	1	Observe the models or obtain an organic model kit and construct models of ethanol, 2-propanol, and *t*-butyl alcohol (2-methyl-2-propanol).	
A	2	Write the condensed structural formula of each.	
A	3	Write the condensed structural formula for phenol.	
A	4	Classify each alcohol as a primary, secondary, or tertiary alcohol.	

B. Properties of Alcohols and Phenol

Materials: 6 test tubes, pH paper, stirring rod, ethanol, 2-propanol, t-butyl alcohol (2-methyl-2-propanol), cyclohexanol, 20% phenol, and unknown

Section	Step	Procedure	Check when finished
B	1	**Odor** Place 5 drops of each of the alcohols, phenol, and unknown to six separate test tubes. *Avoid skin contact with phenol.*	
B	2	Carefully detect the odor of each. Hold your breath as you gently fan some fumes from the top of the test tube toward you as demonstrated by your instructor.	
B	3	Record your observation of the odor.	
B	4	**Solubility in water** Add about 2 mL of water (40 drops) to each test tube.	
B	5	Shake and determine whether each alcohol is soluble or not. If the substance is soluble in water, you will see a clear solution with no separate layers. If it is insoluble, a cloudy mixture or separate layer will form.	
B	6	Record your observations.	

Laboratory 14

C. Oxidation of Alcohols

Materials: 6 test tubes, ethanol, 2-propanol, *t*-butyl alcohol (2-methyl-2-propanol), cyclohexanol, phenol, unknown, 2% chromate solution

Section	Step	Procedure	Check when finished
C.1	1	Place 8 drops of the alcohols and an unknown in separate test tubes.	
C.1	2	Carefully add 2 drops of chromate solution to each. Caution: *Chromate solution contains concentrated H_2SO_4, which is corrosive.*	
C.1	3	Look for a color change in the chromate solution as you add it to the sample. If the orange color turns to green in 1–2 minutes, oxidation of the alcohol has taken place. If the color remains orange, no reaction has occurred.	
C.1	4	If a test tube becomes hot, place it in a beaker of ice-cold water.	
C.1	5	Record your observations.	
C.2	1	Draw the condensed structural formula of each alcohol.	
C.3	1	Classify each alcohol as primary (1°), secondary (2°), or tertiary (3°).	
C.4	1	Draw the condensed structural formulas of the products where oxidation occurred. When there is no change in color, no oxidation took place. Write "no reaction" (NR).	

D. Properties of Aldehydes and Ketones

Materials: Organic model kits, Chemistry handbook, test tubes, droppers, 5- or 10-mL graduated cylinder, acetone, benzaldehyde, camphor, vanillin, cinnamaldehyde, 2,3-butanedione, propionaldehyde, cyclohexanone, and unknown

Section	Step	Procedure	Check when finished
D	1	Make models or observe models of formaldehyde, acetaldehyde, propionaldehyde, acetone, butanone, and cyclohexanone.	
D	1	Draw their condensed structural formulas.	
D	1	Write the IUPAC and common names (if any) for each.	

Odors of Aldehydes and Ketones

Section	Step	Procedure	Check when

Alcohols, Aldehydes, and Ketones

Section	Step	Procedure	Check when finished
D.1	1	Carefully detect the odor of samples of acetone, benzaldehyde, camphor, vanillin, cinnamaldehyde, and 2,3-butanedione, and unknown. Your instructor will demonstrate this technique.	
D.1	2	Record your observations.	
D.2	1	Draw their condensed structural formulas. You may need a chemistry handbook or a *Merck Index*.	
D.2	2	Identify each as a ketone or aldehyde.	

Solubility of Aldehydes and Ketones

Section	Step	Procedure	Check when finished
D.3	1	Place 2 mL of water in each of 5 separate test tubes.	
D.3	2	Add 5 drops of propionaldehyde (propanal), benzaldehyde, acetone, cyclohexanone, and unknown to each tube respectively. Mix by shaking.	
D.3	3	Record your observations. *Save the samples for part C.*	

E. Iodoform Test for Methyl Ketones

Materials: Test tubes from part B.3, dropper, 10% NaOH, warm water bath, and iodine test reagent

Section	Step	Procedure	Check when finished
E	1	Using the test tubes from part B.3, add 10 drops of 10% NaOH to each.	
E	2	Warm the tubes in a warm water bath to 50–60°C.	
E	3	Add 20 drops of iodine test reagent. Mix by shaking.	
E	4	Look for the formation of a yellow solid precipitate.	
E	5	Record your results.	

F. Oxidation of Aldehydes and Ketones

Materials: Test tubes, propionaldehyde (propanal), benzaldehyde, acetone, cyclohexanone, unknown, Benedict's reagent, droppers, boiling water bath

Section	Step	Procedure	Check when finished
F	1	Place 10 drops of propionaldehyde (propanal), benzaldehyde, acetone, cyclohexanone and unknown in separate test tubes.	
F	2	Label.	

Laboratory 14

F	3	Add 2 mL of Benedict's reagent to each test tube. Mix by shaking.	
F	4	Place the test tubes in a boiling water bath for 5 minutes.	
F	5	The appearance of the red-orange color of Cu_2O indicates that oxidation has occurred. Moderate amounts of Cu_2O will blend with the blue Cu^{2+} solution to form green or rust color.	
F	6	Record your observations.	
F	7	Identify the compounds that gave an oxidation reaction.	

Alcohols, Aldehydes, and Ketones

Report Sheet - Lab 14

Date _____ Name _____
Section _____ Team _____
Instructor _____ _____

Pre-Lab Study Questions

1. What is the functional group of an alcohol and a phenol?

2. Why are some alcohols soluble in water?

3. How are alcohols classified?

4. If you added the oxidizing agent, chromate, to a solution of phenol, would the phenol be oxidized and would a green Cr^{3+} solution be formed?

Laboratory 14

5. What is the functional group of an aldehyde? A ketone?

6. What is the oxidation product of an aldehyde?

7. If you add Benedict's reagent to a solution of an aldehyde and a ketone, which solution would remain blue?

Alcohols, Aldehydes, and Ketones

Report Sheet - Lab 14

Date _____ Name _____
Section _____ Team _____
Instructor _____   _____

A. Structures of Alcohols and Phenols

Ethanol	2-Propanol
Classification:	
t-Butyl alcohol (2-methyl-2-propanol)	Phenol
Classification:	

Questions and Problems

Q.1   Write the structures and classifications of the following alcohols:

1-Pentanol	3-Pentanol
Cyclopentanol	1-Methylcyclopentanol

Q.2   In your textbook or a chemistry handbook, look up and draw the condensed structural formulas and uses of thymol, menthol, and resorcinol. Circle the phenol functional group in each structure. If the compounds are available in lab, carefully note and describe their odors.

Laboratory 14

Report Sheet - Lab 14

B. Properties of Alcohols and Phenols

Alcohol	Odor	Soluble in Water?
Ethanol		
2-Propanol		
t-Butyl alcohol		
Cyclohexanol		
Phenol		
Unknown		

Questions and Problems

Q.3   Write the product of the following reactions (if no reaction, write NR):

Report Sheet - Lab 14

C. Oxidation of Alcohols

Alcohol	C.1 Color Change with CrO_4^{2-}	C.2 Condensed Structural Formula	C.3 Classification	C.4 Oxidation Product (If reaction takes place)
Ethanol				
2-Propanol				
t-Butyl alcohol				
Cyclohexanol				
Phenol				
Unknown				

Laboratory 14

Report Sheet - Lab 14

D. Properties of Aldehydes and Ketones

Structures

Formaldehyde IUPAC Name _____	Acetaldehyde IUPAC Name _____
Propionaldehyde IUPAC Name _____	Acetone IUPAC Name _____
Butanone Common Name _____	Cyclohexanone

Odor

	B.1 **Odor**	B.2 **Condensed Structural Formula**	**Aldehyde or Ketone?**
Acetone			
Benzaldehyde			
Camphor			
Vanillin			
Cinnamaldehyde			
2,3-Butanedione			
Unknown		✕	

Report Sheet - Lab 14

Questions and Problems

Q.1 What aldehyde or ketone might be present in the following everyday products?

Artificial butter flavor in popcorn _____

Almond-flavored cookies _____

Candies with cinnamon flavor _____

Nail polish remover _____

D., E., and F. Solubility, Iodoform, and Oxidation of Aldehydes and Ketones

	D.3 Solubility Soluble in water?	E. Iodoform Test Methyl ketone present?	F. Benedict's Test Oxidation occurred?
Propionaldehyde			
Benzaldehyde			
Acetone			
Cyclohexanone			
Unknown			

Questions and Problems

Q.2 Complete the following with the word *soluble or insoluble:*

Aldehydes and ketones containing one to four carbon atoms are _____ in water.

Aldehydes and ketones containing five or more carbon atoms are _____ in water.

Laboratory 14

Report Sheet - Lab 14

Q.3 Indicate the test results for each of the following compounds in the iodoform test and in the Benedict's test:

	Iodoform Test	Benedict's Test
$CH_3CCH_2CH_3$ (with C=O)		
CH_3CH (with C=O)		
$CH_3CH_2CCH_2CH_3$ (with C=O)		
CH_3CCH_2CH (with two C=O)		

Q.4 Two compounds, A and B, have the formula of C_3H_6O. Determine their condensed structural formulas and names using the following test results.

a. Compound A forms a red-orange precipitate with Benedict's reagent but does not react with iodoform.

b. Compound B forms a yellow solid in the iodoform test but does not react with Benedict's reagent.

15

Carboxylic Acids and Esters

Goals

- Write the structural formulas of carboxylic acids and esters.
- Determine the solubility and acidity of carboxylic acids and their salts.
- Write equations for neutralization and esterification of acids.
- Prepare esters and identify their characteristic odors.
- Use an esterification reaction to synthesize aspirin.
- Purify the crude aspirin sample by recrystallization.

Text References

Review the following before performing this laboratory.

Timberlake; *General, Organic, and Biological Chemistry: Structures of Life*, Third Edition
Sections 16.1 to 16.6; Pages 579 to 594

or

Timberlake; *Chemistry: An Introduction to General, Organic, and Biological Chemistry*, Tenth Edition
Sections 13.1 to 13.3; Pages 448 to 460

Discussion

In the 18th century, an extract of willow bark was found useful in reducing fevers (antipyretic) and relieving pain (analgesic) and inflammation (anti-inflammatory). Although salicylic acid was effective at reducing fever and pain, it damaged the mucous membranes of the mouth and esophagus, and caused hemorrhaging of the stomach lining. At the turn of the century, scientists at the Bayer Company in Germany noted that salicylic acid contained a phenol group that might cause the damage. They decided to modify the salicylic acid by forming an ester with a two-carbon acetyl group. The resulting substance was acetylsalicylic acid, or ASA, which we call aspirin. Aspirin acts by inhibiting the formation of prostaglandins, 20-carbon acids that form at the site of an injury and cause inflammation and pain.

In commercial aspirin products, a small amount of acetylsalicylic acid (300 mg to 400 mg) is bound together with a starch binder and sometimes caffeine and buffers to make an aspirin tablet. The basic conditions in the small intestine break down the acetylsalicylic acid to yield salicylic acid, which is absorbed into the bloodstream. The addition of a buffer reduces the irritation caused by the carboxylic acid group of the aspirin molecule.

Laboratory 15

A. Carboxylic Acids and Their Salts

A salad dressing made of oil and vinegar tastes tart because it contains vinegar, which contains acetic acid (ethanoic acid). The sour taste of fruits such as lemons is due to acids such as citric acid. Face creams contain alpha hydroxy acids such as glycolic acid. All these acids are carboxylic acids, which contain the carboxyl group: a carbonyl group attached to a hydroxyl group. A dicarboxylic acid such as malonic acid, found in apples, has two carboxylic acid functional groups. The carboxylic acid of benzene is called benzoic acid.

$$-\overset{O}{\underset{\|}{C}}-OH$$
Carboxyl group

$$CH_3\overset{O}{\underset{\|}{C}}-OH \qquad CH_3CH_2\overset{O}{\underset{\|}{C}}-OH \qquad HO-\overset{O}{\underset{\|}{C}}-CH_2-\overset{O}{\underset{\|}{C}}-OH \qquad C_6H_5-\overset{O}{\underset{\|}{C}}-OH$$
Acetic acid | Propionic acid | Malonic acid | Benzoic acid

Ionization of Carboxylic Acids in Water

Carboxylic acids are weak acids because the carboxylic acid group ionizes slightly in water to give a proton and a carboxylate ion as shown in the equation below. However, like the alcohols, the polarity of the carboxylic acid group makes acids with one to four carbon atoms soluble in water. Acids with two or more carboxyl groups (diacids) are more soluble in water.

$$CH_3-\overset{O}{\underset{\|}{C}}-OH + H_2O \rightleftharpoons CH_3-\overset{O}{\underset{\|}{C}}-O^- + H_3O^+$$
Acetic acid | Acetate ion (a carboxylate ion)

Neutralization of Carboxylic Acids

An important feature of carboxylic acids is their neutralization by bases such as sodium hydroxide to form carboxylate salts and water. We saw in an earlier experiment that neutralization is the reaction of an acid with a base to give a salt and water. Even insoluble carboxylic acids with five or more carbon atoms can be neutralized to give corresponding salts that are usually soluble in water. For this reason, acids used in food products or medications are in their soluble carboxylate salt form rather than the acid.

$$HX + NaOH \longrightarrow Na^+X^- + H_2O$$
Acid | Base | Salt | Water

$$CH_3-\overset{O}{\underset{\|}{C}}-OH + NaOH \longrightarrow CH_3-\overset{O}{\underset{\|}{C}}-O^-Na^+ + H_2O$$
Acetic acid | Sodium acetate (a carboxylate salt)

B. Esters

Carboxylic acids may have tart or unpleasant odors, but many esters have pleasant flavors and fragrant odors. Octyl acetate gives oranges their characteristic odor and flavor; pear flavor is due to pentyl acetate. The flavor and odor of raspberries come from isobutyl formate.

Carboxylic Acids and Esters

$$CH_3(CH_2)_7O-\overset{O}{\underset{\|}{C}}-CH_3$$
Octyl acetate
(oranges)

$$CH_3(CH_2)_4-O-\overset{O}{\underset{\|}{C}}-CH_3$$
Pentyl acetate
(pears)

$$CH_3-\overset{CH_3}{\underset{|}{C}}HCH_2-O-\overset{O}{\underset{\|}{C}}-H$$
Isobutyl formate
(raspberries)

An ester of salicylic acid is methyl salicylate, which gives the flavor and odor of oil of wintergreen used in candies and ointments for sore muscles. When salicylic acid reacts with acetic anhydride, acetylsalicylic acid (ASA) is formed, which is aspirin, widely used to reduce fever and inflammation.

Methyl salicylate
(wintergreen)

Acetylsalicylic acid
(aspirin)

Esterification and Hydrolysis

In a reaction called *esterification,* the carboxylic acid group combines with the hydroxyl group of an alcohol. The reaction, which takes place in the presence of an inorganic acid catalyst, produces an ester and water.

Esterification ⟶

$$CH_3-\overset{O}{\underset{\|}{C}}-OH + HO-(CH_2)_4-CH_3 \; \underset{H^+}{\rightleftharpoons} \; CH_3-\overset{O}{\underset{\|}{C}}-O-(CH_2)_4-CH_3 + H_2O$$

Acetic acid   1-Pentanol   Pentyl acetate (pear flavor)

⟵ Hydrolysis

The reverse reaction, hydrolysis, occurs when an acid catalyst and water cause the decomposition of an ester to yield the carboxylic acid and alcohol. The ester product is favored when an excess of acid or alcohol is used; hydrolysis is favored when more water is present.

C. Preparation of Aspirin

Aspirin (acetylsalicylic acid) can be prepared from acetic acid and the hydroxyl group on salicylic acid. However, this is a slow reaction. The ester forms rapidly when acetic anhydride is used to provide the acetyl group in the presence of acid acting as a catalyst. The aspirin you will prepare in this experiment is impure and must not be taken internally!

Salicylic acid (138 g/mole) + Acetic anhydride $\xrightarrow{H^+}$ Aspirin (acetylsalicylic acid) (180 g/mole) + Acetic acid

Laboratory 15

Using the following equation, the maximum amount (yield) of aspirin that is possible from 2.00 g of salicylic acid can be calculated.

$$2.00 \text{ g salicylic acid} \times \frac{1 \text{ mole salicylic acid}}{138 \text{ g}} \times \frac{1 \text{ mole aspirin}}{1 \text{ mole salicylic acid}} \times \frac{180 \text{ g}}{1 \text{ mole aspirin}}$$

$$= 2.61 \text{ g aspirin (possible)}$$

Suppose the total amount of aspirin you obtain has a mass of 2.25 g. A percentage yield can be calculated as follows:

$$\% \text{ Yield} = \frac{\text{g aspirin obtained}}{\text{g aspirin calculated}} \times 100 = 86.2\% \text{ yield of aspirin product}$$

Lab Information

Time: 2½ hr

Comments: When noting odors, hold your breath and fan across the top of a test tube to detect the odor. The formation of esters requires concentrated acid. Use carefully. Tear out the report sheets and place them beside the matching procedure.

Related Topics: Carboxylic acid functional group, ester functional group, ionization of carboxylic acids, neutralization, esterification, hydrolysis, saponification

Dispose of all chemicals as directed by your laboratory instructor.

Experimental Procedures GOGGLES REQUIRED!

A. Carboxylic Acids and Their Salts

Materials: Test tubes, glacial acetic acid, benzoic acid(s), dropper, spatula, pH paper, red and blue litmus paper, stirring rod, 400-mL beaker, hot plate or Bunsen burner, 10% NaOH, 10% HCl

Section	Step	Procedure	Check when finished
A.1	1	Write the structural formulas for acetic acid and benzoic acid.	

Section	Step	Procedure	Check when finished
A.2	1	Place about 2 mL of water in two test tubes	

Carboxylic Acids and Esters

Section	Step	Procedure	Check when finished
A.2	2	Add 5 drops of acetic acid to one test tube and a small amount of benzoic solid (enough to cover the tip of a spatula) to the other.	
A.2	3	Tap the sides of the test tubes to mix or stir with a stirring rod.	
A.2	4	Identify the acid that dissolves.	

Section	Step	Procedure	Check when finished
A.3	1	Test the pH of each carboxylic acid by dipping a stirring rod into the solution, then touching it to a piece of pH paper.	
A.3	2	Compare the color on the paper with the color chart on the container and report the pH.	

Section	Step	Procedure	Check when finished
A.4	1	Place the test tube of benzoic acid (solid should be present) in a hot water bath and heat for 5 minutes.	
A.4	2	Describe the effect of heating on the solubility of acid.	
A.4	3	Allow the test tube to cool.	
A.4	4	Record your observations.	

Section	Step	Procedure	Check when finished
A.5	1	Add about 10 drops of 10% NaOH to each test tube until a drop of the solution turns red litmus paper blue.	
A.5	2	Record your observations.	
A.5	3	Write the equations for the reactions with NaOH including the structures of the sodium salts formed.	

Section	Step	Procedure	Check when finished
A.6	1	Add about 10 drops of 10% HCl to each sample until it is neutralized (blue litmus paper turns red).	
A.6	2	Record your observations.	
A.6	3	Write equations for the reactions of the sodium salts that formed.	

B. Esters

Materials: Organic model set, test tubes, hot plate or Bunsen burner, 400-mL beaker, stirring rod, spatula, small beaker, methanol, 1-pentanol, 1-octanol, benzyl alcohol, 1-propanol, salicylic acid(s), glacial acetic acid, 85% H_3PO_4

Laboratory 15

Section	Step	Procedure	Check when finished
B.1	1	Make a model of acetic acid and methyl alcohol.	
B.1	2	Remove the components of water and form an ester bond to give methyl acetate.	
B.1	3	Write the equation for the formation of the ester.	

Section	Step	Procedure	Check when finished
B.2	1	As assigned, prepare one of the mixtures listed by placing 3 mL of the alcohol in a test tube and label it with the mixture number. **Mixture Alcohol Carboxylic Acid** 1 Methanol Salicylic acid 2 1-Pentanol Glacial acetic acid 3 1-Octanol Glacial acetic acid 4 Benzyl alcohol Glacial acetic acid 5 1-Propanol Glacial acetic acid *Caution: Use care in dispensing glacial acetic acid. It can cause burns and blisters on the skin.*	
B.2	2	Add 2 mL of the assigned carboxylic acid or the amount of solid that covers the tip of a spatula.	
B.2	3	Write the condensed structural formulas for the alcohols and carboxylic acids.	

Section	Step	Procedure	Check when finished
B.3	1	With the test tube of the sample prepared in B.2 pointed away from you, *cautiously* add 15 drops of concentrated phosphoric acid, H_3PO_4.	
B.3	2	Stir by shaking or with a stirring rod.	
B.3	3	Place the test tube in a boiling water bath for 15 minutes.	
B.3	4	Remove the test tube and *cautiously* fan the vapors toward you as previously demonstrated by your instructor.	
B.3	5	Record the odors you detect such as pear, banana, orange, raspberry, or oil of wintergreen (spearmint).	
B.3	6	For a stronger odor, place 15 mL of hot water in a small beaker and pour the ester solution into the hot water.	

Carboxylic Acids and Esters

Section	Step	Procedure	Check when finished
B.3	7	If specified by your instructor, repeat the esterification with other mixtures of alcohol and carboxylic acid. Note the odors of esters produced by other students.	
B.3	8	Write the condensed structural formulas and names of the esters produced.	
B.3	9	*Dispose of the ester products as instructed.*	

C. Preparation of Aspirin

Materials: 125-mL Erlenmeyer flask, 400-mL beaker, hot plate or Bunsen burner, ice, salicylic acid, acetic anhydride, 5- or 10-mL graduated cylinder, stirring rod, pan or large beaker, dropper, 85% H_3PO_4 in a dropper bottle, Büchner filtration apparatus, filter paper, spatula, watch glass

Section	Step	Procedure	Check when finished
C.1	1	Weigh a 125-mL Erlenmeyer flask.	
C.1	2	Record as C.1.	
C.1	3	Add 2.00 g of salicylic acid and reweigh.	
C.1	4	Record as C.1a.	
C.1	5	*Working in the hood, carefully* add 5 mL of acetic anhydride to the flask. *Caution: Acetic anhydride is irritating to the nose and sinuses. Handle carefully.*	
C.1	6	Slowly add 10 drops of 85% phosphoric acid, H_3PO_4.	
C.1	7	Stir the mixture with a stirring rod.	
C.1	8	Place the flask and its contents in a boiling water bath and stir until all the solid dissolves.	
C.1	9	Remove the flask from the hot water and let it cool to room temperature.	
C.1	10	*Working in the hood, cautiously* add 20 drops of water to the cooled mixture. ***KEEP YOUR FACE AWAY FROM THE TOP OF THE FLASK: ACETIC ACID VAPORS ARE IRRITATING***	
C.1	11	When the reaction is complete (when no further change is observed), add 50 mL of cold water.	
C.1	12	Cool the mixture by placing the flask in an ice bath for 10 minutes.	
C.1	13	Stir. Crystals of aspirin should form.	

C.1	14	If no crystals appear, gently scratch the inside of the flask with a stirring rod.	
C.1	15	**Collecting the Aspirin Crystals: See Figure 15.1.** Add a piece of filter paper to a Büchner filtration apparatus.	
C.1	16	Place the funnel in the filter flask making sure that the neck fits snugly in a rubber washer.	
C.1	17	Moisten the filter paper.	
C.1	18	Turn on the water aspirator and pour the aspirin product onto the filter paper in the Büchner funnel.	
C.1	19	Push down gently on the funnel to create the suction needed to pull the water off the aspirin product.	
C.1	20	The aspirin crystals will collect on the filter paper. See Figure 15.1.	
C.1	21	Use a spatula to transfer any crystals left in the flask.	
C.1	22	Rinse the inside of the flask with a 10-mL portion of ice cold water to transfer all the crystals to the funnel.	
C.1	23	Wash the aspirin crystals on the filter paper with two 10-mL portions of cold water.	
C.1	24	Spread the aspirin crystals out on the filter paper and draw air through the funnel. This helps dry the crystals.	
C.1	25	Turn off the water and use a spatula to lift and transfer the filter paper and aspirin to a paper towel. Don't touch it; it may still contain acid.	
C.1	26	Allow the crystals to air dry.	

Figure 15.1 Apparatus for suction filtration with a Büchner funnel

Carboxylic Acids and Esters

Section	Step	Procedure	Check when finished
C.2	1	**Calculations:** Weigh a clean, dry watch glass.	
C.2	2	Record as C.2.	
C.3	1	Transfer the acetylsalicylic acid crystals to the watch glass.	
C.3	2	Weigh the watch glass and the aspirin. (C.3)	
C.4	1	Calculate the mass of salicylic acid. (C.1 − C.1a))	
C.5	1	Calculate the maximum yield of aspirin possible from the salicylic acid. (C.4 × 180)/138	
C.6	1	Calculate the mass of the aspirin you collected. (C.3 − C.2)	
C.7	1	Calculate the percent yield of aspirin.(C.6/C.5) × 100	
C.8	1	If a melting point apparatus is available, determine the melting point of your aspirin product. Pure aspirin has a melting point of 135°C. Salicylic acid melts at 157–159°C. Compare the melting point of your aspirin with the known melting points of aspirin and salicylic acid. Your instructor will demonstrate the use of this apparatus.	

Laboratory 15

Report Sheet - Lab 15

Date _____ Name _____
Section _____ Team _____
Instructor _____   _____

Pre-Lab Study Questions

1. What are some carboxylic acids you encounter in daily life?

2. What are esters?

3. What functional group of aspirin causes it to irritate the stomach?

Carboxylic Acids and Esters

Report Sheet - Lab 15

Date _____ Name _____
Section _____ Team _____
Instructor _____   _____

A. Carboxylic Acids and Their Salts

	Acetic Acid	Benzoic Acid
A.1 Condensed structural formulas		
A.2 Solubility in cold water		
A.3 pH		
A.4 Solubility in hot water	✗	
A.5 NaOH		
A.6 HCl		

Write equations for the reactions of the carboxylic acids with NaOH (A.5).

Write equations for the reactions of the salts of the carboxylic acids with HCl (A.6).

Questions and Problems

Q.1 How does NaOH affect the solubility of benzoic acid in water? Why?

Laboratory 15

Report Sheet - Lab 22

Q.2 Write the names of the following carboxylic acid and ester:

a. $CH_3CH_2CH_2-\overset{\overset{O}{\|}}{C}-OH$

b. $CH_3CH_2\overset{\overset{O}{\|}}{C}-OCH_3$

Q.3 Why are there differences in the solubility of the carboxylic acids in part A?

B. Esters

B.1 Equation for the formation of methyl acetate:

B.2 Condensed Structural Formulas of Alcohol and Carboxylic Acid	B.3 Odor of Ester	Condensed Structural Formula and Name of Ester
Methanol and salicylic acid		
1-Pentanol and acetic acid		
1-Octanol and acetic acid		
Benzyl alcohol and acetic acid		
1-Propanol and acetic acid		

Report Sheet - Lab 15

A. Preparation of Aspirin

C.1   Mass of flask   _____

C.1a Mass of flask and salicylic acid   _____

C.2   Mass of watch glass   _____

C.3   Mass of watch glass and crude aspirin product   _____

Calculations

C.4   Mass of salicylic acid   _____

C.5   Possible (maximum) yield of aspirin   _____
      (*Show calculations.*)

C.6   Mass of prepared aspirin   _____

C.7   Percent yield   _____
      (*Show calculations.*)

C.8   Melting point (°C) of aspirin product (optional)   _____

Laboratory 15

Report Sheet - Lab 15

Questions and Problems

Q.4 Write the structural formula for aspirin. Label the ester group and the carboxylic acid group.

Q.5 If a typical aspirin tablet contains 325 mg aspirin (the rest is starch binder), how many tablets could you prepare from the aspirin you made in lab?

16
Carbohydrates

Goals

- Identify the characteristic functional groups of carbohydrates.
- Describe common carbohydrates and their sources.
- Distinguish between monosaccharides, disaccharides, and polysaccharides.
- Observe physical and chemical properties of some common carbohydrates.
- Use physical and chemical tests to distinguish between monosaccharides, disaccharides, and polysaccharides.
- Identify an unknown carbohydrate.
- Relate the process of digestion to the hydrolysis of carbohydrates.

Text References

Review the following before performing this laboratory.

Timberlake; *General, Organic, and Biological Chemistry: Structures of Life*, Third Edition
Sections 14.5 and 15.1 to 15.6; Pages 525 to 532 and 543 to 566

or

Timberlake; *Chemistry: An Introduction to General, Organic, and Biological Chemistry*, Tenth Edition
Sections 12.6 and 14.1 to 14.6 ; Pages 427 to 432 and 483 to 506

Discussion

Carbohydrates in our diet are our major source of energy. Foods high in carbohydrates include potatoes, bread, pasta, and rice. If we take in more carbohydrates than we need for energy, the excess is converted to fat, which can lead to a weight gain. The carbohydrate family can be organized into three classes, which are the monosaccharides, disaccharides, and polysaccharides.

A. Monosaccharides

Monosaccharides contain C, H, and O in units of $(CH_2O)_n$. Most common monosaccharides have six carbon atoms (hexoses) with a general formula of $C_6H_{12}O_6$. They contain many hydroxyl groups (–OH) along with a carbonyl group. The aldoses are monosaccharides with an aldehyde group, and ketoses contain a ketone group in open chain form.

Laboratory 16

Monosaccharides		Sources
Glucose	$C_6H_{12}O_6$	Fruit juices, honey, corn syrup
Galactose	$C_6H_{12}O_6$	Lactose hydrolysis
Fructose	$C_6H_{12}O_6$	Fruit juices, honey, sucrose hydrolysis

Glucose, a hexose, is the most common monosaccharide; it is also known as blood sugar.

```
        O
        ‖
        C—H
        |
    H—C—OH
        |
   HO—C—H
        |
    H—C—OH
        |
    H—C—OH
        |
        CH₂OH
      D-Glucose
```

The letter D refers to the orientation of the hydroxyl group on the chiral carbon that is farthest from the carbonyl group at the top of the chain (carbon 1). The D- and L-isomers of glyceraldehyde illustrate the position of the –OH on the central, chiral atom.

```
    D-isomer                L-isomer
       H   O                   H   O
        \ //                    \ //
         C                       C
         |                       |
    H—C—OH                 HO—C—H
         |                       |
       CH₂OH                   CH₂OH
   D-Glyceraldehyde        L-Glyceraldehyde
```

Haworth Structures

Most of the time glucose exists in a ring structure, which forms when the OH on carbon 5 forms a hemiacetal bond with the aldehyde group on carbon. In the Haworth structure the new hydroxyl group on carbon 1 may be drawn above carbon 1 (the β form) or below carbon 1 (the α form).

Carbohydrates

B. Disaccharides

A disaccharide contains two monosaccharides bonded together. Some common disaccharides include maltose, sucrose (table sugar), and lactose (milk sugar).

Disaccharides	Sources	Monosaccharides
Maltose	Germinating grains, starch hydrolysis	Glucose + glucose
Lactose	Milk, yogurt, ice cream	Glucose + galactose
Sucrose	Sugar cane, sugar beets	Glucose + fructose

In a disaccharide, two monosaccharides form a glycosidic bond with the loss of water. For example, in maltose, two glucose units are linked by an α-1,4-glycosidic bond.

C. Polysaccharides

Polysaccharides are long-chain polymers that contain many thousands of monosaccharides (usually glucose units) joined together by glycosidic bonds. Three important polysaccharides are starch, cellulose, and glycogen. They all contain glucose units, but differ in the type of glycosidic bonds and the amount of branching in the polymer.

Laboratory 16

Polysaccharides	Found in	Monosaccharides
Starch (amylose, amylopectin)	Rice, wheat, grains, cereals	Glucose
Glycogen	Muscle, liver	Glucose
Cellulose	Wood, plants, paper, cotton	Glucose

Starch is an insoluble storage form of glucose found in rice, wheat, potatoes, beans, and cereals. Starch is composed of two kinds of polysaccharides, amylose and amylopectin. *Amylose*, which makes up about 20% of starch, consists of α-D-glucose molecules connected by α-1,4-glycosidic bonds in a continuous chain. A typical polymer of amylose may contain from 250 to 4000 glucose units.

α-1,4-Glycosidic bonds in amylose

Amylopectin is a branched-chain polysaccharide that makes up as much as 80% of starch. In amylopectin, α-1,4-glycosidic bonds connect most of the glucose molecules. However, at about every 25 glucose units, there are branches of glucose molecules attached by α-1,6-glycosidic bonds between carbon 1 of the branch and carbon 6 in the main chain.

Amylopectin

Glycogen is a similar to amylopectin but it is even more highly branched, with α-1,6-glycosidic bonds about every 10 to 15 glucose units.

Carbohydrates

Cellulose is the major structural material of wood and plants. Cotton is almost pure cellulose. In cellulose, glucose molecules form a long unbranched chain similar to amylose except that β-1,4-glycosidic bonds connect the glucose molecules. The β isomers are aligned in parallel rows that are held in place by hydrogen bonds between the rows. This gives a rigid structure for cell walls in wood and fiber and makes cellulose more resistant to hydrolysis.

Cellulose

D. Benedict's Test for Reducing Sugars

All of the monosaccharides and most of the disaccharides can be oxidized to carboxylic acids. When the cyclic structure opens, the aldehyde group is available for oxidation. Benedict's reagent contains Cu^{2+} ion that is reduced. Therefore, all the sugars that react with Benedict's reagent are called *reducing sugars*. Ketoses also act as reducing sugars because the ketone group on carbon 2 isomerizes to give an aldehyde group on carbon 1.

Ketose ⇌ Aldose (can be oxidized)

When oxidation of a sugar occurs, the Cu^{2+} is reduced to Cu^+, which forms a red precipitate of cuprous oxide, $Cu_2O(s)$. The color of the precipitate varies from green to gold to red depending on the concentration of the reducing sugar.

sugar—CH(=O) + 2Cu^{2+} →[OH^-] sugar—COH(=O) + $Cu_2O(s)$

Reducing sugar | Blue | Oxidized sugar | Red

Sucrose is not a reducing sugar because it cannot revert to the open-chain form that would provide the aldehyde group needed to reduce the cupric ion.

Laboratory 16

[Structure of sucrose showing 1,2-Glycosidic bond; annotation: "These carbons cannot open. Sucrose is not a reducing sugar."]

Sucrose

E. Seliwanoff's Test for Ketoses

Seliwanoff's test is used to distinguish between hexoses with a ketone group (ketoses) and hexoses that are aldehydes (aldoses). With ketoses, a deep red color is formed rapidly. Aldoses give a light pink color that takes a longer time to develop. The test is most sensitive for fructose, which is a ketose.

F. Fermentation Test

Most monosaccharides and disaccharides undergo fermentation in the presence of yeast. The products of fermentation are ethyl alcohol (CH_3CH_2OH) and carbon dioxide (CO_2). The formation of bubbles of carbon dioxide is used to confirm the fermentation process.

$$C_6H_{12}O_6 \xrightarrow{yeast} 2C_2H_5OH + 2CO_2(g)$$
Glucose　　　　　　Ethanol

Although enzymes are present for the hydrolysis of most disaccharides, they are not available for lactose. The enzymes needed for the fermentation of galactose are not present in yeast. Therefore, lactose and galactose give negative results with the fermentation test.

G. Iodine Test for Polysaccharides

When iodine (I_2) is added to amylose, the helical shape of the unbranched polysaccharide traps iodine molecules, producing a deep blue-black complex. Amylopectin, and cellulose, react with iodine to give red to brown colors. Glycogen produces a reddish-purple color. Monosaccharides and disaccharides are too small to trap iodine molecules and do not form dark colors with iodine.

H. Hydrolysis of Disaccharides and Polysaccharides

Disaccharides hydrolyze in the presence of an acid to give the individual monosaccharides.

$$Sucrose + H_2O \xrightarrow{H^+} Glucose + Fructose$$

Carbohydrates

In the laboratory, we use water and acid to hydrolyze starches, which produce smaller saccharides such as maltose. Eventually, the hydrolysis reaction converts maltose to glucose molecules. In the body, enzymes in our saliva and from the pancreas carry out the hydrolysis. Complete hydrolysis produces glucose, which provides about 50% of our nutritional calories.

$$\text{Amylose, amylopectin} \xrightarrow{H^+ \text{ or amylase}} \text{dextrins} \xrightarrow{H^+ \text{ or amylase}} \text{maltose} \xrightarrow{H^+ \text{ or maltase}} \text{many D-glucose units}$$

Lab Information

Time:	2 hr
Comments:	Tear out the report sheets and place them next to the matching procedures. In the study of carbohydrates, it is helpful to review stereoisomers and the formation of hemiacetals,
Related Topics:	Carbohydrates, monosaccharides, disaccharides, polysaccharides, hemiacetals, stereoisomers, aldohexoses, ketohexoses, chiral compounds, Fischer projection, Haworth structures, reducing sugars, fermentation

Experimental Procedures

A. Monosaccharides

Materials: Organic model kits or prepared models

Section	Step	Procedure	Check when finished
A.1	1	Make or observe models of L-glyceraldehyde and D-glyceraldehyde.	
A.1	2	Draw the Fischer projections.	
A.2	1	Draw the Fischer projection for D-glucose.	
A.2	2	Draw the Haworth (cyclic) formulas for the α and β anomers for D-glucose.	
A.3	1	Draw the Fischer projections for D-fructose and D-galactose.	
A.3	2	Draw the Haworth (cyclic) formulas for the α anomers of each.	

B. Disaccharides

Section	Step	Procedure	Check when finished
B.1	1	Using Haworth formulas, write the structure for α-D-maltose. Look at a model if available.	

Laboratory 16

Section	Step	Procedure	Check when finished
B.2	1	Write an equation for the hydrolysis of α-D-maltose by adding H_2O to the glycosidic bond.	
B.3	1	Using Haworth formulas, write an equation for the formation of α-D-lactose from β-D-galactose and α-D-glucose.	
B.4	1	Draw the structure of sucrose and circle the glycosidic bond.	

C. Polysaccharides

Section	Step	Procedure	Check when finished
C.1	1	Draw a portion of amylose using four units of α-D-glucose. Indicate the glycosidic bonds.	
C.2	1	Describe how the structure of amylopectin differs from the structure of amylose.	
C.3	1	Draw a portion of cellulose using four units of β-D-glucose. Indicate the glycosidic bonds.	

D. Benedict's Test for Reducing Sugars

Materials: Test tubes, 400-mL beaker, droppers, hot plate or Bunsen burner, 5- or 10-mL graduated cylinder, Benedict's reagent, 2% carbohydrate solutions: glucose, fructose, sucrose, lactose, starch, and an unknown

Section	Step	Procedure	Check when finished
D	1	Place 10 drops of solutions of glucose, fructose, sucrose, lactose, starch, water, and unknown in separate test tubes.	
D	2	Label each test tube.	
D	3	Add 2 mL of Benedict's reagent to each sample and mix by shaking.	
D	4	Place the test tubes in a boiling water bath for 3–4 minutes. The formation of a greenish to reddish-orange color indicates the presence of a reducing sugar. If the solution is the same color as the Benedict's reagent in water (the control), there has been no oxidation reaction.	
D	5	Record your observations.	
D	6	Classify each as a reducing or nonreducing sugar.	

Carbohydrates

E. Seliwanoff's Test for Ketoses

Materials: Test tubes, 400-mL beaker, droppers, hot plate or Bunsen burner, 5- or 10-mL graduated cylinder, Seliwanoff's reagent, 2% carbohydrate solutions: glucose, fructose, sucrose, lactose, starch, and an unknown

Section	Step	Procedure	Check when finished
E	1	Place 10 drops of solutions of glucose, fructose, sucrose, lactose, starch, water, and unknown in separate test tubes.	
E	2	Label each test tube.	
E	3	Add 2 mL of Seliwanoff's reagent to each. Mix by shaking. *The reagent contains concentrated HCl. Use carefully.*	
E	4	Place the test tubes in a boiling hot water bath and note the time.	
E	5	After 1 minute, observe the colors in the test tubes. A rapid formation of a deep red color indicates the presence of a ketose.	
E	6	Record your results as a fast color change, slow change, or no change.	

F. Fermentation Test

Materials: Fermentation tubes (or small and large test tubes), baker's yeast, 2% carbohydrate solutions: glucose, fructose, sucrose, lactose, starch, and an unknown

Section	Step	Procedure	Check when finished
F	1	Fill fermentation tubes with a solution of glucose, fructose, sucrose, lactose, starch, water, and unknown.	
F	2	Add 0.2 g of yeast to each and mix well by shaking. See Figure 16.1.	
F	3	If fermentation tubes are not available, use small test tubes placed upside down in larger test tubes.	
F	4	Cover the mouth of the large test tube with filter paper or cardboard.	
F	5	Place your hand firmly over the paper cover and invert	
F	6	When the small test tube inside has completely filled with the mixture, return the larger test tube to an upright position. See Figure 16.2.	
F	7	Set the tubes aside.	

Laboratory 16

F	8	At the end of the laboratory period, and again at the next laboratory period, look for gas bubbles in the fermentation tubes or inside the small tubes.	
F	9	Record your observations. See Figure 16.3.	

FIGURE 16.1 Fermentation tube filled with a carbohydrate solution

FIGURE 16.2 Test tubes used as fermentation tubes

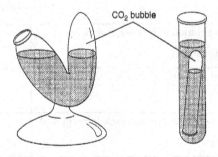

FIGURE 16.3 Fermentation tubes with CO_2 bubbles

Carbohydrates

G. Iodine Test for Polysaccharides

Materials: Spot plate or test tubes, droppers, iodine reagent, 2% carbohydrate solutions in dropper bottles: glucose, fructose, sucrose, lactose, starch, and an unknown

Section	Step	Procedure	Check when finished
G	1	Using a spot plate, place 5 drops of each solution of glucose, fructose, sucrose, lactose, starch, water, and unknown in the wells. (If you do not have a spot plate, use small test tubes.)	
G	2	Add 1 drop of iodine solution to each sample. A dark blue-black color is a positive test for amylose in starch. A red or brown color indicates the presence of other polysaccharides.	
G	3	Record your results.	
G	4	Complete the table to identify your unknown.	

H. Hydrolysis of Disaccharides and Polysaccharides

Materials: Test tubes, 10-mL graduated cylinder, 400-mL beaker (boiling water bath), hot plate or Bunsen burner, spot plate or watch glass, 10% HCl, 10% NaOH, red litmus paper, iodine reagent, Benedict's reagent, 2% starch and sucrose solutions in dropper bottles

Section	Step	Procedure	Check when finished
H	1	Place 3 mL of 2% starch in two test tubes.	
H	2	Place 3 mL of 2% sucrose in two more test tubes.	
H	3	To one sample each of sucrose and starch, add 20 drops of 10% HCl. Mix by shaking.	
H	4	To the other samples of sucrose and starch, add 20 drops of H_2O. Mix by shaking.	
H	5	Label the test tubes and heat in a boiling water bath for 10 minutes.	
H	6	Remove the test tubes from the water bath and let them cool.	
H	7	To the samples containing HCl, add 10% NaOH (about 20 drops) until one drop of the mixture turns litmus paper blue, indicating the HCl has been neutralized.	
H	8	Test separate samples for hydrolysis using the Iodine Test and Benedict's Test as described below.	
H	9	**Iodine Test** Place 5 drops of each solution on a spot plate or watch glass.	
H	10	Add 1 drop of iodine reagent to each.	
H	11	Record observations.	

Laboratory 16

H	12	Determine if hydrolysis has occurred in each.	
H	13	**Benedict's Test** Add 2 mL of Benedict's reagent to each of the samples.	
H	14	Heat each sample in a boiling water bath for 3–4 minutes.	
H	15	Determine if hydrolysis has occurred in each.	

Carbohydrates

Report Sheet - Lab 16

Date _____ Name _____
Section _____ Team _____
Instructor _____   _____

Pre-Lab Study Questions

1. What are some sources of carbohydrates in your diet?

2. What does the D in D-glucose mean?

3. What is the bond that links monosaccharides in di- and polysaccharides?

4. What happens to glucose or galactose when the Cu^{2+} in Benedict's reagent is reduced?

5. Would you expect fructose or glucose to form a red color rapidly with Seliwanoff's reagent?

Laboratory 16

6. Why don't all the disaccharides undergo fermentation with yeast?

7. How can the iodine test be used to distinguish between amylose and glycogen?

Carbohydrates

Report Sheet - Lab 16

Date _____ Name _____
Section _____ Team _____
Instructor _____   _____

A. Monosaccharides

A.1 Fischer projections

L-glyceraldehyde   D-glyceraldehyde

How does L-glyceraldehyde differ from D-glyceraldehyde?

A.2 Fischer projection of D-glucose   Haworth (cyclic) formulas

α-D-glucose β-D-glucose

A.3 Fischer projection of D-fructose   Haworth (cyclic) formula for α-D-fructose

Laboratory 16

Report Sheet - Lab 16

Fischer projection of D-galactose   Haworth (cyclic) formula for α-D-galactose

Questions and Problems

Q.1   How does the structure of D-glucose compare to the structure of D-galactose?

B.   Disaccharides

B.1   Draw the structure of α-D-maltose

B.2   Equation for the hydrolysis of α-D-maltose

B.3   Equation for the formation of α-D-lactose

Report Sheet - Lab 16

B.4 Draw the structure of sucrose.

Questions and Problems

Q.2 What is the type of glycosidic bond in maltose?

Q.3 Why does maltose have both α and β anomers? Explain.

C. Polysaccharides

C.1 Draw a portion of amylose.

C.2 Comparison of amylopectin to amylose.

Laboratory 16

Report Sheet - Lab 16

C.3 Draw a portion of cellulose.

Questions and Problems

Q.4 What is the monosaccharide that results from the complete hydrolysis of amylose?

Q.5 What is the difference in the structure of amylose and cellulose?

Results of Carbohydrate Tests

	D. Benedict's Test	E. Seliwanoff's Test	F. Fermentation Test	G. Iodine Test
Glucose				
Fructose				
Sucrose				
Lactose				
Starch				
Water				
Unknown # _____				

Carbohydrates

Report Sheet - Lab 16

Questions and Problems

Q.6 From the results in part D, list the sugars that are reducing sugars and those that are not.
Reducing sugars

Nonreducing sugars

Q.7 What sugars are ketoses?

Q.8 What sugars give a positive fermentation test?

Q.9 Which carbohydrates give a blue-black color in the iodine test?

Laboratory 16

Report Sheet - Lab 16

Identifying an Unknown Carbohydrate

Unknown No._____

	Results with Unknown	Possible Sugars Present
Benedict's (D)		
Seliwanoff's (E)		
Fermentation (F)		
Iodine (G)		

What carbohydrate(s) is/are in your unknown?

Questions and Problems

Q.10 What carbohydrate(s) would have the following test results?

a. Produces a reddish-orange solid with Benedict's and a red color with Seliwanoff's reagent in 1 minute

b. Gives a color change with Benedict's test, a light orange color with Seliwanoff's reagent after 5 minutes, and produces no bubbles during fermentation

Report Sheet - Lab 16

    c.   Gives no color change with Benedict's or Seliwanoff's test, but turns a blue-black color with iodine reagent

H. Hydrolysis of Disaccharides and Polysaccharides

Results	Sucrose + H$_2$O	Sucrose + HCl	Starch + H$_2$O	Starch + HCl
Iodine test				
Benedict's test				
Hydrolysis products present				

Questions and Problems

Q.11   How do the results of the Benedict's test indicate that hydrolysis of sucrose and starch occurred?

Q.12   How do the results of the iodine test indicate that hydrolysis of starch occurred?

Laboratory 16

Q.13   Indicate whether the following carbohydrates will give a positive (+) or a negative (-) result in each type of test listed below:

	Benedict's Test	Seliwanoff's Test	Fermentation Test	Iodine Test
Glucose				
Fructose				
Galactose				
Sucrose				
Lactose				
Maltose				
Amylose				
Amylopectin				

17

Lipids

Goals

- Observe the physical and chemical properties of some common lipids.
- Draw the structure of a typical triacylglycerol.
- Distinguish between saturated and unsaturated fats.
- Determine the degree of unsaturation of some fats.
- Prepare a hand lotion and determine the function of its components.

Text References

Review the following before performing this laboratory.

Timberlake; *General, Organic, and Biological Chemistry: Structures of Life*, Third Edition
Sections 17.1 to 17.4; Pages 603 to 620

or

Timberlake; *Chemistry: An Introduction to General, Organic, and Biological Chemistry*, Tenth Edition
Sections 15.1 to 15.4; Pages 514 to 530

Discussion

A. Triacylglycerols

The triacylglycerols, commonly called fats or oils, are esters of glycerol and fatty acids. Fatty acids are long-chain carboxylic acids, usually 14 to 18 carbons in length. When the fatty acid contains carbon to carbon double bonds, the triacylglycerol is referred to as an unsaturated fat. When the fatty acid consists of only carbon to carbon single bonds, the triacylglycerol is a saturated fat. Table 17.1 gives the formulas of the common fatty acids and their melting points. At room temperature, saturated fats are usually solid and unsaturated fats are usually liquid. Fats that contain mostly saturated fatty acids have a higher melting point than fats with more unsaturated fatty acids.

$$\begin{array}{c}
CH_2-OH \\
| \\
CH-OH \\
| \\
CH_2-OH
\end{array}
\;+\;
\begin{array}{c}
O \\
\| \\
HO-C(CH_2)_{14}CH_3 \\
O \\
\| \\
HO-C(CH_2)_{14}CH_3 \\
O \\
\| \\
HO-C(CH_2)_{14}CH_3
\end{array}
\;\longrightarrow\;
\begin{array}{c}
O \\
\| \\
CH_2-O-C(CH_2)_{14}CH_3 \\
O \\
\| \\
CH-O-C(CH_2)_{14}CH_3 \\
O \\
\| \\
CH_2-O-C(CH_2)_{14}CH_3
\end{array}
\;+\; 3\,H_2O$$

Glycerol   3 Palmitic acids   Glyceryl palmitate (tripalmitin)

Table 17.1 *Formulas, Melting Points, and Sources of Some Fatty Acids*

# C Atoms	Structural Formula	Melting Point (°C)	Common Name	Source
Saturated fatty acids (single carbon–carbon bonds)				
12	$CH_3(CH_2)_{10}COOH$	44	lauric	coconut
14	$CH_3(CH_2)_{12}COOH$	54	myristic	nutmeg
16	$CH_3(CH_2)_{14}COOH$	63	palmitic	palm
18	$CH_3(CH_2)_{16}COOH$	70	stearic	animal fat
Monounsaturated fatty acids (one cis double bond)				
16	$CH_3(CH_2)_5CH=CH(CH_2)_7COOH$	1	palmitoleic	butter fat
18	$CH_3(CH_2)_7CH=CH(CH_2)_7COOH$	4	oleic	olives, corn
Polyunsaturated fatty acids (two or more cis double bonds)				
18	$CH_3(CH_2)_4CH=CHCH_2CH=CH(CH_2)_7COOH$	−5	linoleic	safflower, sunflower
18	$CH_3CH_2CH=CHCH_2CH=CHCH_2CH=CH(CH_2)_7COOH$	−11	linolenic	corn

B. Physical Properties of Some Lipids and Fatty Acids

Lipids are a family of compounds that are grouped by similarities in solubility rather than structure. As a group, lipids are more soluble in nonpolar solvents such as ether, chloroform, or benzene than in water. Most are not soluble in water. Important types of lipids include fats and oils, glycerophospholipids, and steroids. Compounds classified as lipids include fat-soluble vitamins A, D, E, and K; cholesterol; hormones; portions of cell membranes; and vegetable oils. Table 17.2 lists the classes of lipids.

Lipids

Table 17.2 *Classes of Lipid Molecules*

Lipids	Components
Waxes	Fatty acid and long-chain alcohol
Fats and oils (triacylglycerols)	Fatty acids and glycerol
Glycerophospholipids	Fatty acids, glycerol, phosphate, amino alcohol
Sphingolipids	Fatty acids, sphingosine, phosphate, amino alcohol
Glycosphingolipids	Fatty acids, glycerophospholipids sphingosine, monosaccharides
Steroids	A fused structure of three cyclohexanes and a cyclopentane

The structural formulas of three typical lipids are shown below:

$$CH_3-(CH_2)_{14}-\overset{O}{\underset{\|}{C}}-O-(CH_2)_{29}-CH_3$$
Wax

Triacylglycerol, a fat

Cholesterol, a steroid

C. Bromine Test for Unsaturation

The presence of unsaturation in a fatty acid or a triacylglycerol can be detected by the bromine test, which you used in an earlier experiment to detect double bonds in alkenes. If the orange color of the bromine solution fades quickly, an addition reaction has occurred and the oil or fat is unsaturated.

Laboratory 17

Bromine adds to the double bond

$$CH_2-O-\overset{O}{\overset{\|}{C}}(CH_2)_7CH=CH(CH_2)_7CH_3$$
$$|$$
$$CH-O-\overset{O}{\overset{\|}{C}}(CH_2)_{16}CH_3 \quad + \text{Br}_2 \longrightarrow$$
$$|$$
$$CH_2-O-\overset{O}{\overset{\|}{C}}(CH_2)_{16}CH_3$$

$$CH_2-O-\overset{O}{\overset{\|}{C}}(CH_2)_7\overset{Br}{\overset{|}{C}}H-\overset{Br}{\overset{|}{C}}H(CH_2)_7CH_3$$
$$|$$
$$CH-O-\overset{O}{\overset{\|}{C}}(CH_2)_{16}CH_3$$
$$|$$
$$CH_2-O-\overset{O}{\overset{\|}{C}}(CH_2)_{16}CH_3$$

D. Preparation of Hand Lotion

We use hand lotions and creams to soften our skin and reduce dryness. Typically, the formulation of a hand lotion consists of several components such as stearic acid, lanolin, triethanolamine, cetyl alcohol, glycerin (glycerol), water, and usually a fragrance. Lanolin from wool consists of a mixture of waxes.

Cetyl alcohol $CH_3(CH_2)_{15}OH$

Stearic acid $CH_3-(CH_2)_{16}-\overset{O}{\overset{\|}{C}}-OH$

Glycerol (glycerine)
$$CH_2-OH$$
$$|$$
$$CH-OH$$
$$|$$
$$CH_2-OH$$

Triethanolamine
$$CH_2CH_2OH$$
$$|$$
$$HOCH_2CH_2-N-CH_2CH_2OH$$

Because lipids are nonpolar, they protect and soften by preventing the loss of moisture from the skin. Some of the components help emulsify the polar and nonpolar ingredients. In this experiment, we will see how the physical and chemical properties of lipids are used to prepare a hand lotion.

Lab Information

Time: 3 hr

Comments: *Bromine can cause severe chemical burns. Use carefully.*

Tear out the Lab report sheets and place them beside the matching procedures.

Related topics: Fatty acids, saturated and unsaturated fatty acids, lipids, triglycerides

Dispose of all chemicals as directed by your laboratory instructor.

Experimental Procedures   *GOGGLES REQUIRED!*

Lipids

A. Triacylglycerols

Materials: Organic model kits or models

Section	Step	Procedure	Check when finished
A.1	1	Use an organic model kit or observe prepared models of a molecule of glycerol and three molecules of ethanoic acid.	
A.1	2	What are the functional groups on each?	
A.1	3	Draw their structures.	
A.2	1	Form ester bonds between the hydroxy groups on glycerol and the carboxylic acid groups of the ethanoic acid molecules. In the process, three molecules of water are removed.	
A.2	2	Write an equation for the formation of the glyceryl ethanoate.	
A.3	1	Carry out the reverse process, which is hydrolysis. Add the components of water to break the ester bond.	
A.3	2	Add an arrow to the equation to show the reverse direction for the hydrolysis reaction.	

B. Physical Properties of Some Lipids and Fatty Acids

Materials: Test tubes and stoppers, dropper bottles or solids: stearic acid, oleic acid, olive oil, safflower oil, lecithin, cholesterol, vitamin A capsules, spatulas, CH_2Cl_2 (optional)

Section	Step	Procedure	Check when finished
B	1	**Appearance and Odor:** To seven separate test tubes, add 5 drops or the amount of solid lipid held on the tip of a spatula: stearic acid, oleic acid, olive oil, safflower oil, lecithin, cholesterol, vitamin A (puncture a capsule or use cod liver oil).	
B.1	1	Classify each as a triacylglycerol (fat or oil), fatty acid, steroid, or phospholipid.	
B.2	1	Describe their appearance.	
B.3	1	Describe their odors.	

Section	Step	Procedure	Check when finished
B.4	1	**Solubility in a Polar Solvent** *(May be a demonstration)*: Add about 2 mL of water to each of the test tubes.	

Laboratory 17

B.4	2	Stopper and shake each test tube.	
B.4	3	Record your observations.	

Section	Step	Procedure	Check when finished
B.5	1	**Solubility in Non Polar Solvent** *(May be done as a demonstration.)*: To seven separate test tubes, add 5 drops or the amount of solid lipid held on the tip of a spatula: stearic acid, oleic acid, olive oil, safflower oil, lecithin, cholesterol, vitamin A (puncture a capsule or use cod liver oil).	
B.5	2	Add 1 mL (20 drops) methylene chloride, CH_2Cl_2, to each sample.	
B.5	3	Record the solubility of the lipids. *Save the test tubes and samples of stearic acid, oleic acid, olive oil and safflower oil for part C.*	

C. Bromine Test for Unsaturation

Materials: Samples from B.5, 1% Br_2 in methylene chloride

Section	Step	Procedure	Check when finished
C	1	To the samples from B.5, add 1% bromine solution drop by drop until a permanent red-orange color is obtained or until 20 drops have been added. *Caution: Avoid contact with bromine solution; it can cause painful burns. Do not breathe the fumes.*	
C	2	Record your observations.	
C	3	Determine if the red-orange color fades rapidly or persists.	

D. Preparation of Hand Lotion

Materials: Stearic acid, cetyl alcohol, lanolin (anhydrous), triethanolamine, glycerol, ethanol, distilled water, fragrance (optional), commercial hand lotions, 10-mL graduated cylinder, 50-mL graduated cylinder, 50-mL or 100-mL beakers, thermometer, 100-mL beakers, 250-mL beaker for water bath, Bunsen burner, iron ring, wire screen, stirring rods, tongs, pH paper

Team project: D.1, D.2, and D.3 may be prepared by different teams in the lab.

Lipids

Section	Step	Procedure	Check when finished
D.1	1	Obtain the following substances and combine in two 50-mL or 100-mL beakers. Use a laboratory balance to weigh out the solid substances. Use a 10-mL graduated cylinder to measure small volumes, and a 50-mL graduated cylinder to measure larger volumes.	
D.1	2	**Beaker 1** 　　Stearic acid　　　　　3 g 　　Cetyl alcohol　　　　 1 g 　　Lanolin (anhydrous)　 2 g 　　Triethanolamine　　　1 mL	
D.1	3	**Beaker 2** 　　Glycerin　　　　　　 2 mL 　　Water　　　　　　　 50 mL	
D.1	4	**Water bath:** Fill a 250-mL beaker about 2/3 full of water. Place the beaker on a hot plate. (Alternatively, an iron ring covered with a wire screen with a Bunsen burner under it can be used.) Heat the water.	
D.1	5	Using a pair of crucible tongs, hold Beaker 1 (four ingredients) in the water bath and heat to 80°C or until all the compounds have melted.	
D.1	6	Remove from the water bath.	
D.1	7	Using a pair of crucible tongs, place Beaker 2 (two ingredients) in the same water bath and heat to 80°C.	
D.1	8	While still warm, slowly pour the glycerol-water mixture from Beaker 2 into Beaker 1 (four ingredients) as you stir.	
D.1	9	Add 5.0 mL of ethanol and a few drops of fragrance, if desired.	
D.1	10	Continue to stir for 3 to 5 minutes until a smooth, creamy lotion is obtained. If the resulting product is too thick, add more warm water.	
D.1	11	Describe the smoothness and appearance of the hand lotion.	

Laboratory 17

Section	Step	Procedure	Check when finished
D.2	1	**Omitting triethanolamine from the product:** Obtain the following substances and combine in two 50-mL or 100-mL beakers. Use a laboratory balance to weigh out the solid substances. Use a 10-mL graduated cylinder to measure small volumes, and a 50-mL graduated cylinder to measure larger volumes.	

D.2	2	**Beaker 1**
		    Stearic acid 3 g
		    Cetyl alcohol 1 g
		    Lanolin (anhydrous) 2 g	

D.2	3	**Beaker 2**
		    Glycerin 2 mL
		    Water 50 mL	

D.2	4	**Water bath:** Fill a 250-mL beaker about 2/3 full of water. Place the beaker on a hot plate. (Alternatively, an iron ring covered with a wire screen with a Bunsen burner under it can be used.) Heat the water.	
D.2	5	Using a pair of crucible tongs, hold Beaker 1 (four ingredients) in the water bath and heat to 80°C or until all the compounds have melted.	
D.2	6	Remove from the water bath.	
D.2	7	Using a pair of crucible tongs, place Beaker 2 (two ingredients) in the same water bath and heat to 80°C.	
D.2	8	While still warm, slowly pour the glycerol-water mixture from Beaker 2 into Beaker 1 (four ingredients) as you stir.	
D.2	9	Add 5.0 mL of ethanol and a few drops of fragrance, if desired.	
D.2	10	Continue to stir for 3 to 5 minutes until a smooth, creamy lotion is obtained. If the resulting product is too thick, add more warm water.	
D.2	11	Describe the smoothness and appearance of the hand lotion.	
D.2	12	Compare the properties of the resulting hand lotion to the one obtained in D.1 and to commercial hand lotions.	

Lipids

Section	Step	Procedure	Check when finished
D.3	1	**Omitting stearic acid from the product:** Obtain the following substances and combine in two 50-mL or 100-mL beakers. Use a laboratory balance to weigh out the solid substances. Use a 10-mL graduated cylinder to measure small volumes, and a 50-mL graduated cylinder to measure larger volumes.	

D.3	2	**Beaker 1**
		     Cetyl alcohol 1 g
		     Lanolin (anhydrous) 2 g
		     Triethanolamine 1 mL	

D.3	3	**Beaker 2**
		     Glycerin 2 mL
		     Water 50 mL	

D.3	4	**Water bath:** Fill a 250-mL beaker about 2/3 full of water. Place the beaker on a hot plate. (Alternatively, an iron ring covered with a wire screen with a Bunsen burner under it can be used.) Heat the water.	
D.3	5	Using a pair of crucible tongs, hold Beaker 1 (four ingredients) in the water bath and heat to 80°C or until all the compounds have melted.	
D.3	6	Remove from the water bath.	
D.3	7	Using a pair of crucible tongs, place Beaker 2 (two ingredients) in the same water bath and heat to 80°C.	
D.3	8	While still warm, slowly pour the glycerol-water mixture from Beaker 2 into Beaker 1 (four ingredients) as you stir.	
D.3	9	Add 5.0 mL of ethanol and a few drops of fragrance, if desired.	
D.3	10	Continue to stir for 3 to 5 minutes until a smooth, creamy lotion is obtained. If the resulting product is too thick, add more warm water.	
D.3	11	Describe the smoothness and appearance of the hand lotion.	
D.3	12	Compare the properties of the resulting hand lotion to the one obtained in D.1 and to commercial hand lotions.	

Section	Step	Procedure	Check when finished
D.4	1	Determine the pH of the hand lotions you and others in your lab have prepared along with any commercial hand lotions.	

Laboratory 17

Lipids

Report Sheet - Lab 17

Date _____ Name _____
Section _____ Team _____
Instructor _____ _____

Pre-Lab Study Questions

1. What is the functional group in a triacylglycerol?

2. Write the structure of linolenic acid. Why is it an unsaturated fatty acid?

Laboratory 17

Lipids

Report Sheet - Lab 17

Date _____ Name _____
Section _____ Team _____
Instructor _____   _____

A. Triacylglycerols

A.1 Structure of glycerol   Structure of ethanoic acid

A.2 Reversible equation for the esterification and hydrolysis of glyceryl ethanoate

B. Physical Properties of Some Lipids and Fatty Acids

Lipid	B.1 Type of Lipid	B.2 Appearance	B.3 Odor	B.4 Soluble in Water?	B.5 Soluble in CH_2Cl_2?
Stearic acid					
Oleic acid					
Olive oil					
Safflower oil					
Lecithin					
Cholesterol					
Vitamin A					

Laboratory 17

Report Sheet - Lab 17

Questions and Problems

Q.1 Why are the compounds in part B classified as lipids?

Q.2 What type of solvent is needed to remove an oil spot? Why?

C. Bromine Test for Unsaturation

Compound	Drops of Bromine Solution	Saturated or Unsaturated?
Fatty acids		
Stearic acid		
Oleic acid		
Triacylglycerols		
Safflower oil		
Olive oil		

Questions and Problems

Q.3 a. Write the condensed structural formulas of stearic acid and oleic acid.

Stearic acid

Oleic acid

Report Sheet - Lab 17

b. Which fatty acid is unsaturated?

c. The melting point of stearic acid is 70°C, and that of oleic acid is 4°C. Explain the difference.

d. From the results of experiment C, how can you tell which is more unsaturated, oleic acid or stearic acid?

D. Preparation of Hand Lotion

Descriptions of the hand lotions		
D.1	D.2 (without triethanolamine)	D.3 (without stearic acid)
D.4 pH of the hand lotions		
pH of commercial hand lotions		
Brand_____	Brand_____	Brand_____
pH _____	pH _____	pH _____

Laboratory 17

Report Sheet - Lab 17

Questions and Problems

Q. 4   How does omitting triethanolamine affect the properties and appearance of the hand lotion?

Q. 5   How does omitting stearic acid affect the properties and appearance of the hand lotion?

Q. 6   What would be a reason to have triethanolamine and stearic acid as ingredients in hand lotion?

18

Saponification and Soaps

Goals

- Prepare soap by the saponification of a fat or oil.
- Observe the reactions of soap with oil, CaCl₂, MgCl₂, and FeCl₃.

Text References

Review the following before performing this laboratory.

Timberlake; *General, Organic, and Biological Chemistry: Structures of Life*, Third Edition
Section 17.4; Pages 615 to 624

or

Timberlake; *Chemistry: An Introduction to General, Organic, and Biological Chemistry*, Tenth Edition
Section 15.4; Pages 525 to 530

Discussion

A. Saponification

When an ester is hydrolyzed in the presence of a base, the reaction is called saponification. The products are the salt of the carboxylic acid and the alcohol. Although the ester is usually insoluble in water, the salt and alcohol (if short-chain) are soluble.

$$CH_3-\underset{\text{Ethyl acetate (ester)}}{\overset{\overset{O}{\|}}{C}-O-CH_2CH_3} + NaOH \longrightarrow CH_3-\underset{\text{Sodium acetate (carboxylate salt)}}{\overset{\overset{O}{\|}}{C}-O^-Na^+} + \underset{\text{Ethanol (alcohol)}}{HO-CH_2CH_3}$$

(Ester bond splits)

B. Saponification: Preparation of Soap

For centuries, soaps have been made from animal fats and lye (NaOH), which was obtained by pouring water through wood ashes. The hydrolysis of a fat or oil by a base such as NaOH is called *saponification* and the salts of the fatty acids obtained are called *soaps*. The other product of hydrolysis is glycerol, which is soluble in water.

Laboratory 18

$$CH_2-O-\overset{O}{\underset{\|}{C}}-(CH_2)_{14}CH_3$$
$$CH-O-\overset{O}{\underset{\|}{C}}-(CH_2)_{14}CH_3 \quad + \quad 3NaOH \quad \longrightarrow \quad CH-OH \quad + \quad 3Na^+ \quad {}^-O\overset{O}{\underset{\|}{C}}-(CH_2)_{14}CH_3$$
$$CH_2-O-\overset{O}{\underset{\|}{C}}-(CH_2)_{14}CH_3$$

| Fat (tripalmitin) | Base | Glycerol | Soap (sodium palmitate) |

The fats that are most commonly used to make soap are lard and tallow from animal fat and coconut, palm, and olive oils from vegetables. Castile soap is made from olive oil. Soaps that float have air pockets. Soft soaps are made with KOH instead of NaOH to give potassium salts.

C. Properties of Soaps and Detergents

A soap molecule has a dual nature. The nonpolar carbon chain is hydrophobic and attracted to nonpolar substances such as grease. The polar head of the carboxylate salt is hydrophilic and attracted to water.

The dual polarity of a soap (salt of a fatty acid)

$$CH_3CH_2CH_2CH_2CH_2CH_2CH_2CH_2CH_2CH_2CH_2CH_2CH_2CH_2CH_2CH_2CH_2 - \overset{O}{\underset{\|}{C}} - O^-Na^+$$

Nonpolar tail (hydrophobic) — Polar head (hydrophilic)

When soap is added to a greasy substance, the hydrophobic tails are embedded in the nonpolar fats and oils. However, the polar heads are attracted to the polar water molecules. Clusters of soap particles called *micelles* form with the nonpolar oil droplet in the center surrounded by many polar heads that extend into the water. Eventually all of the greasy substance forms micelles, which can be washed away with water. In hard water (high concentration of minerals), the carboxylate ends of soap react with Ca^{2+}, Fe^{3+}, or Mg^{2+} ions and form an insoluble substance, which we see as a gray line in the bathtub or sink. Tests will be done with the soap you prepare to measure its pH, its ability to form suds in soft and hard water, and its reaction with oils.

Detergents or "syndets" are called synthetic cleaning agents because they are not derived from naturally occurring fats or oils. They are popular because they do not form insoluble salts with ions, which means they work in hard water as well as in soft water. A typical detergent is sodium lauryl sulfate.

$$CH_3(CH_2)_{10}CH_2-O-\overset{O}{\underset{\underset{O}{\|}}{S}}-O^-Na^+$$

Lauryl sulfate salt, a nonbiodegradable detergent

As detergents replaced soaps for cleaning, it was found that they were not degraded in sewage treatment plants. Large amounts of foam appeared in streams and lakes that became polluted with detergents.

Biodegradable detergents such as an alkylbenzenesulfonate detergent eventually replaced the nonbiodegradable detergents.

$$CH_3(CH_2)_9-CH(CH_3)-C_6H_4-S(=O)_2-O^-Na^+$$

Laurylbenzenesulfonate salt, a biodegradable detergent

In addition to the sulfonate salts, a box of detergent contains phosphate compounds along with brighteners and perfumes. However, phosphates accelerate the growth of algae in lakes and cause a decrease in the dissolved oxygen in the water. As a result, the lake eutrophies. Some replacements for phosphates have been made.

Lab Information

Time: 3 hr

Comments: *You will be working with hot oil and NaOH. Be sure you wear your goggles.* Tear out the report sheets and place them beside the matching procedures.

Related Topics: Esters, saponification, soaps, hydrophobic, hydrophilic

Dispose of all chemicals as directed by your laboratory instructor.

Experimental Procedures GOGGLES REQUIRED!

A. Saponification

Materials: Test tube, test tube holder, droppers, stirring rod, hot plate or Bunsen burner, 250- or 400mL beaker, methyl salicylate in a dropper bottle, 10% NaOH, 10% HCl, blue litmus paper

Section	Step	Procedure	Check when finished
A.1	1	Draw the condensed structural formula of methyl salicylate.	

Section	Step	Procedure	Check when finished
A.2	1	Place 3 mL of water in a test tube	
A.2	2	Add 5 drops of methyl salicylate to the test tube.	

Laboratory 18

Section	Step	Procedure	Check when finished
A.2	3	Record the appearance and odor of the ester and save the test tube for A.3	

Section	Step	Procedure	Check when finished
A.3	1	Prepare a boiling water bath using the 250- or 400-mL beaker.	
A.3	2	Add 1 mL (20 drops) of 10% NaOH to the test tube from A.2	
A.3	3	Place the test tube in a boiling water bath for 30 minutes or until the top layer of the ester disappears. There should be two layers in the test tube.	
A.3	4	Record any changes in the odor and appearance of the ester.	
A.3	5	Remove the test tube and cool in cold water. Save this for A.5	

Section	Step	Procedure	Check when finished
A.4	1	Write the equation for the saponification reaction.	

Section	Step	Procedure	Check when finished
A.5	1	After the solution from A.3 is cool, add about 20 drops (1 mL) of 10% HCl until a drop of the solution turns blue litmus paper red.	
A.5	2	Record your observations.	
A.5	3	Determine the formula of the solid that forms.	

B. Saponification: Preparation of Soap

Materials: 150-mL beaker, hot plate, graduated cylinder, stirring rod or stirring hot plate with stirring bar, large watch glass, 400-mL beaker, Büchner filter system, filter paper, plastic gloves, fat (lard, solid shortening, coconut oil, olive or other vegetable oil), ethanol, 20% NaOH, saturated NaCl solution

Section	Step	Procedure	Check when finished
B	1	Weigh a 150-mL beaker.	
B	2	Add about 5 g of fat or oil.	
B	3	Reweigh.	
B	4	Add 15 mL ethanol (solvent) and 15 mL of 20% NaOH. *Use care when pouring NaOH.*	

Saponification and Soaps

B	5	Place the beaker on a hot plate, cover with a watch glass and heat to a gentle boil and stir continuously. A magnetic stirring bar may be used with a magnetic stirrer.	
B	6	Heat for 30 minutes or until saponification is complete and the solution becomes clear with no separation of layers. Be careful of splattering; the mixture contains a strong base. Wear disposable gloves, if available. Do not let the mixture overheat or char.	
B	7	Add 5-mL portions of an ethanol–water (1:1) mixture to maintain volume. If foaming is excessive, *reduce* the heat.	
B		*Caution: Oil and ethanol will be hot, and may splatter or catch fire. Keep a watch glass nearby to smother any flames. NaOH is caustic and can cause permanent eye damage. Wear goggles at all times.*	
B	8	Obtain 50 mL of a saturated NaCl solution in a 400-mL beaker. (A saturated NaCl solution is prepared by mixing 30 g of NaCl with 100 mL of water.)	
B	9	Pour the soap solution into this salt solution and stir. This process, known as "salting out," causes the soap to separate out and float on the surface.	
B	10	Collect the solid soap using a Büchner funnel and filter paper. See Figure 18.1.	
B	11	Wash the soap with two 10-mL portions of cold water.	
B	12	Pull air through the product to dry it further.	
B	13	Place the soap curds on a watch glass or in a small beaker and dry the soap until the next lab session. Use disposable, plastic gloves to handle the soap. **Handle with care: The soap may still contain NaOH, which can irritate the skin.** Save the soap you prepared for the next part of this experiment.	
B	14	Describe the soap.	

Laboratory 18

Figure 18.1 Apparatus for suction filtration with Büchner funnel

C. Properties of Soaps and Detergents

Materials: Test tubes, stoppers to fit, droppers, small beakers, 50- or 100-mL graduated cylinder, stirring rod, laboratory-prepared soap (from part A), commercial soap product, detergent, pH paper, oil, 1% $CaCl_2$, 1% $MgCl_2$, and 1% $FeCl_3$

Section	Step	Procedure	Check when finished
C	1	Prepare solutions of the soap you made in part A, a commercial soap, and a detergent by dissolving 1 g of each in 50 mL of distilled water. If the soap is a liquid, use 20 drops.	
C.1	1	**pH test** Place 10 mL of each soap solution in separate test tubes. Use 10 mL of water as a comparison.	
C.1	2	Label.	
C.1	3	Dip a stirring rod into each solution, rinsing and drying it between solutions to prevent contamination, then touch the stirring rod to pH paper. Determine the pH. *Save the tubes for part B.2.*	
C.2	1	**Foam test** Stopper each of the tubes from B.1.	
C.2	2	Shake for 10 seconds. The soap should form a layer of suds or foam.	
C.2	3	Record your observations. *Save the tubes for part B.3.*	

C.3	1	**Reaction with oil** Add 5 drops of oil to each test tube from B.2.	
C.3	2	Stopper and shake each one for 10 seconds.	
C.3	3	Record your observations. Compare the sudsy layer in each test tube to the sudsy layers in part B.2.	
C.4	1	**Hard water test** Place 5 mL of the soap solutions in three separate test tubes.	
C.4	2	Add 20 drops of 1% $CaCl_2$ to the first sample.	
C.4	3	Add 20 drops of 1% $MgCl_2$ to the second tube.	
C.4	4	Add 20 drops of 1% $FeCl_3$ to the third tube.	
C.4	5	Stopper each test tube and shake 10 seconds.	
C.4	6	Compare the foamy layer in each of the test tubes to the sudsy layer obtained in part B.2.	
C.4	7	Record your observations.	

Report Sheet - Lab 18

Date _____ Name _____
Section _____ Team _____
Instructor _____   _____

Pre-Lab Study Questions

1. What happens when a fatty acid is reacted with NaOH?

2. Why is ethanol added to the reaction mixture of fat and base in the making of soap?

3. Why is the product of saponification a salt?

Laboratory 18

Report Sheet - Lab 18

Date _____ Name _____
Section _____ Team _____
Instructor _____

A. Saponification

A.1 Condensed structural formula of methyl salicylate	
A.2 Appearance and odor of methyl salicylate	
A.3 Describe the appearance and odor of the ester after adding NaOH and heating	
A.4 Write the equation for the saponification of the ester	
A.5 What changes occur when HCl is added?	
What is the formula of the compound that formed when HCl was added?	

B. Saponification: Preparation of Soap

Describe the appearance of your soap.

Report Sheet - Lab 18

Questions and Problems

Q.1 How would soaps made from vegetable oils differ from soaps made from animal fat?

Q.2 How does soap remove an oil spot?

C. Properties of Soaps and Detergents

Tests	Water	Lab Soap	Commercial Soap	Detergent
C.1 pH				
C.2 Foam				
C.3 Oil				
C.4 $CaCl_2$				
$MgCl_2$				
$FeCl_3$				

Questions and Problems

Q.3 Which of the solutions were basic? Why?

Q.4 Write an equation for the saponification of trimyristin with KOH.

19

Amines and Amino Acids

Goals

- Draw the structural formulas and give the names of amines.
- Observe some physical properties of amines and amides.
- Write an equation for the formation of an amine salt.
- Use R groups to determine if an amino acid will be acidic, basic, or neutral; hydrophobic or hydrophilic.
- Use chromatography to separate and identify amino acids.
- Calculate R_f values for amino acids.

Text References

Review the following before performing this laboratory.

Timberlake; *General, Organic, and Biological Chemistry: Structures of Life*, Third Edition
Sections 18.1 to 18.5 and 19.1 to 19.2; Pages 645 to 664 and 673 to 680

or

Timberlake; *Chemistry: An Introduction to General, Organic, and Biological Chemistry*, Tenth Edition
Sections 13.4 to 13.5 and 16.2 to 16.3; Pages 461 to 473 and 554 to 558

Discussion

Amines are considered as derivatives of ammonia in which one or more hydrogen atoms are replaced with alkyl or aromatic groups. The number of alkyl groups attached to the nitrogen atom determines the classification of primary, secondary, or tertiary amines.

NH_3
Ammonia

CH_3-NH-H
Methylamine
(primary, 1°)

$CH_3-NH-CH_3$
Dimethylamine
(secondary, 2°)

$CH_3-N(CH_3)-CH_3$
Trimethylamine
(tertiary, 3°)

Amines are often found as part of compounds that are physiologically active or used in medications.

Neo-Synephrine

Histamine

Methamphetamine (methedrine)

Laboratory 19

A. Solubility of Amines in Water

In water, ammonia and amines with one to four carbon atoms act as weak bases because the unshared pair of electrons on the nitrogen atom attracts protons. The products are an ammonium ion or alkyl ammonium ion and a hydroxide ion.

$$NH_3 + H_2O \longrightarrow NH_4^+ + OH^-$$
Ammonia → Ammonium ion + Hydroxide ion

$$CH_3-NH_2 + H_2O \longrightarrow CH_3-NH_3^+ + OH^-$$
Methylamine → Methylammonium ion + Hydroxide ion

B. Neutralization of Amines with Acids

Because amines are basic, they react with acids to form the amine salt. These amine salts are much more soluble in water than the corresponding amines.

$$CH_3-NH_2 + HCl \longrightarrow CH_3-NH_3^+ + Cl^-$$
Methylamine → Methylammonium chloride (amine salt)

C. Amino Acids

In our body, amino acids are used to build tissues, enzymes, skin, and hair. About half of the naturally occurring amino acids, the *essential amino acids,* must be obtained from the proteins in the diet because the body cannot synthesize them. Amino acids are similar in structure because each has an amino group (–NH$_2$) and a carboxylic acid group (–COOH). Individual amino acids have different organic groups *(R groups)* attached to the alpha (α) carbon atom. Variations in the R groups determine whether an amino acid is hydrophilic or hydrophobic, and acidic, basic, or neutral.

Some R groups contain carbon and hydrogen atoms only, which makes the amino acids nonpolar and hydrophobic ("water-fearing"). Other R groups contain OH or SH atoms and provide a polar area that makes the amino acids soluble in water; they are hydrophilic ("water-loving"). Other hydrophilic amino acids contain R groups that are carboxylic acids (acidic) or amino groups (basic). The R groups of some amino acids used in this experiment are given in Table 19.1.

TABLE 19.1 AMINO ACIDS FOUND IN NATURE

R Group	Amino Acid	Symbol	Polarity	Reaction to Water
H—	Glycine	Gly	Nonpolar	Hydrophobic
CH_3—	Alanine	Ala	Nonpolar	Hydrophobic
C_6H_5-CH_2-	Phenylalanine	Phe	Nonpolar	Hydrophobic
HO—CH_2—	Serine	Ser	Polar	Hydrophilic
HOOC—CH_2—	Aspartic acid	Asp	Polar, acidic	Hydrophilic
HOOC—CH_2—CH_2—	Glutamic acid	Glu	Polar, acidic	Hydrophilic
$H_2NCH_2CH_2CH_2CH_2$—	Lysine	Lys	Polar, basic	Hydrophilic

Ionization of Amino Acids

An amino acid can ionize when the carboxyl group donates a proton, and when the lone pair of electrons on the amino group accepts a proton. Then the carboxyl group has a negative charge, and the amino group has a positive charge. This ionized form of an amino acid, called a *zwitterion* or *dipolar ion,* has a net zero charge.

$$NH_2-CH(CH_3)-C(=O)-OH \quad\quad {}^+NH_3-CH(CH_3)-C(=O)-O^-$$
Alanine   Zwitterion of alanine

In acidic solutions (low pH), the zwitterion *accepts* a proton (H^+) to form an ion with a positive charge. When placed in a basic solution (high pH), the zwitterion *donates* a proton (H^+) to form an ion with a negative charge. This is illustrated using alanine.

Laboratory 19

$$NH_2-\underset{\underset{H}{|}}{\overset{\overset{CH_3}{|}}{C}}H-\overset{\overset{O}{\|}}{C}-O^- \xleftarrow{\text{Base}} \overset{+}{NH_3}-\underset{\underset{H}{|}}{\overset{\overset{CH_3}{|}}{C}}H-\overset{\overset{O}{\|}}{C}-O^- \xrightarrow{\text{Acid}} \overset{+}{NH_3}-\underset{\underset{H}{|}}{\overset{\overset{CH_3}{|}}{C}}H-\overset{\overset{O}{\|}}{C}-OH$$

High pH ⟵ Donates H+ ⟵ Zwitterion (neutral pH) ⟶ accepts H+ ⟶ Low pH
(charge = 1−)   (charge = 0)   (charge = 1+)

D. Chromatography of Amino Acids

Chromatography is used to separate and identify the amino acids in a mixture. Small amounts of amino acids and unknowns are placed along one edge of Whatman #1 paper. The paper is then placed in a container with solvent. With the paper acting like a wick, the solvent flows up the chromatogram, carrying amino acids with it. Amino acids that are more soluble in the solvent will move higher on the paper. Those amino acids that are more attracted to the paper will remain closer to the origin line. After removing and drying the paper, the amino acids can be detected (visualized) by spraying the dried chromatogram with ninhydrin.

The distance each amino acid travels up the paper from the origin (starting line) is measured and the R_f values calculated. The R_f value is the distance traveled by an amino acid compared to the distance traveled by the solvent. See Figure 19.1.

$$R_f = \frac{\text{distance traveled by an amino acid}}{\text{distance traveled by the solvent}}$$

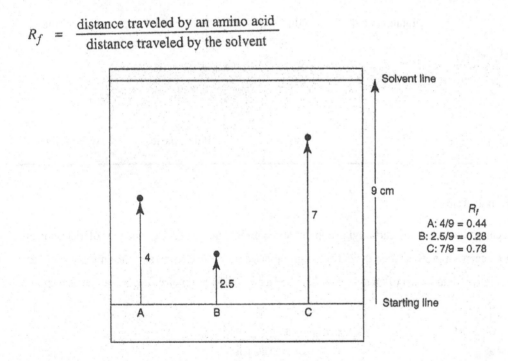

FIGURE 19.1 A developed chromatogram (R_f values calculated for A, B, and C)

To identify an unknown amino acid, its R_f value and color with ninhydrin are compared to the R_f values and colors of known amino acids in the same solvent system. In this way, the amino acids present in an unknown mixture of amino acids can be separated and identified.

Lab Information

Time: 3 hr

Comments: *Some amines have an irritating odor. Work in the hood.* Tear out the report sheets and place them next to the matching procedures.

Related topics: Amines, solubility and pH of amines, amidation, amides, hydrolysis of amides

Dispose of all chemicals as directed by your laboratory instructor.

Experimental Procedures *GOGGLES REQUIRED!*

A. Solubility of Amines in Water

Materials: Aniline, *N*-methylaniline, triethylamine, test tubes, test tube rack, stirring rod, pH paper

WORK IN THE HOOD. THE VAPORS OF AMINES ARE IRRITATING TO THE NOSE AND SINUSES.

Section	Step	Procedure	Check when finished
A.1	1	To three separate test tubes, add 5 drops of aniline, N-methylaniline, and triethylamine.	
A.1	2	Draw the condensed structural formulas of each amine and state its classification (1°, 2°, 3°).	
A.2	1	Cautiously note the odor of each. Remember to hold a fresh breath of air while you fan the vapor toward you. The instructor will demonstrate this technique.	
A.2	2	Record.	
A.3	1	Add 2 mL of water to each test tube.	
A.3	2	Stir with stirring rod, cleaning and drying between test tubes.	
A.3	3	Describe their solubility in water.	
A.4	1	Determine the pH of each solution by dipping a stirring rod in each amine solution and then touching it to pH paper.	
A.4	2	Record.	
A.4	3	*Save these test tubes and samples for part C.*	

B. Neutralization of Amines with Acids

Materials: Test tubes from part B, blue litmus paper, 10% HCl

Laboratory 19

Section	Step	Procedure	Check when finished
B.1	1	Add 10% HCl dropwise to the amine solution until the solution is acidic to litmus paper.	
B.1	2	Record any changes in solubility of each amine.	
B.1	3	Note any changes in odor.	
B.2	1	Write and balance equations for the neutralization of aniline, N-methylaniline, and triethylamine with HCl.	

C. Amino Acids

Materials: Organic model kits or prepared models

Section	Step	Procedure	Check when finished
C.1	1	Using an organic model kit, construct models of glycine and alanine.	
C.1	2	Draw their structures.	
C.1	3	Convert the alanine model to a model of serine.	
C.1	4	Indicate whether each of the amino acids would be hydrophobic or hydrophilic.	
C.2	1	Form ionized (zwitterion) glycine by removing a H atom from the –COOH group and placing it on the N atom in the –NH$_2$ group.	
C.2	2	Draw the structure of the glycine zwitterion.	
C.3	1	Write the structure of glycine in a base and in an acid.	

D. Chromatography of Amino Acids

Materials: 600-mL beaker, plastic wrap, plastic gloves, Whatman chromatography paper #1 (12 cm × 24 cm) or appropriate thin layer chromatographic (TLC) plates, toothpicks or capillary tubing, drying oven (80°C) or hair dryer, metric ruler, stapler, amino acids (1% solutions): phenylalanine, alanine, glutamic acid, serine, lysine, aspartic acid, and unknown

Chromatography solvent: isopropyl alcohol, 0.5 M NH$_4$OH; 0.2% ninhydrin spray

Make all markings on the TLC plate or paper in pencil.

Section	Step	Procedure	Check when finished
D	1	Using forceps or plastic gloves (or a sandwich bag), pick up a piece of Whatman #1 chromatography paper that has been cut to a	

		size of 12 cm × 24 cm. *Keep your fingers off the paper because amino acids can be transferred from the skin.* When this paper is rolled into a cylinder, it should fit into the chromatography tank (large beaker) without touching the sides.
D	1a	Obtain a TLC plate.
D	2	Draw a pencil (lead) line about 2 cm from the long edge of the paper or plate. This will be the starting or origin line.
D	3	With a pencil, mark off seven equally-spaced points about 3 cm apart along the line. (See Figure 19.2.)
D	4	Place your name or initials in the upper corner with the pencil.

FIGURE 19.2 Preparation of a chromatogram

Section	Step	Procedure	Check when finished
D	5	Use the toothpick applicators or capillary tubes provided in each amino acid solution to make a small spot (the size of the letter o) by lightly touching the tip to the paper or plate. *Always return the applicator to the same amino acid solution.*	
D	6	Apply small amounts of a 1% amino acid solution of phenylalanine and label the spot.	
D	7	Apply small amounts of a 1% amino acid solution of glutamic acid and label the spot.	
D	8	Apply small amounts of a 1% amino acid solution of serine . and label the spot.	

Laboratory 19

Section	Step	Procedure	Check when finished		
D	9	Apply small amounts of a 1% amino acid solution of lysine and label the spot.			
D	10	Apply small amounts of a 1% amino acid solution of aspartic acid . and label the spot.			
D	11	Apply small amounts of a 1% amino acid solution of unknown and label the spot.			
D	12	Dry the spot either with a hair dryer or by allowing it to air dry.			
D	13	After the spot dries, repeat steps 6 through 12 twice more.			

Section	Step	Procedure	Check when finished
D	14	Prepare the solvent by mixing 10 mL of 0.5 M NH_4OH and 20 mL of isopropyl alcohol.	
D	15	Pour the solvent into a 600-mL beaker to a depth of about 1 cm but not over 1.5 cm. (The height of the solvent must not exceed the height of the origin line on your chromatography paper.)	
D	16	Cover the beaker tightly with plastic wrap. This is your chromatography tank.	
D	17	Label the beaker with your name and leave it in the hood.	

Section	Step	Procedure	Check when finished
D	18	If using paper, roll the paper into a cylinder and staple the edges *without overlapping. The edges should not touch. See Fig 19.3. If using a plate, skip this step.*	
D	19	Slowly lower the cylinder into the solvent of the chromatography tank with the row of amino acids at the bottom. Make sure that the paper does not touch the sides of the beaker. See Figure 30.3. Or put the plate into the solvent.	
D	20	Cover the beaker with the plastic wrap and leave it undisturbed. The tank must not be disturbed while solvent flows up the paper or plate.	
D	21	Let the solvent rise until it is 2–3 cm from the top edge of the paper. It may take 45–60 minutes. *Do not let the solvent run over the top of the paper.*	

FIGURE 19.3 Chromatogram in a solvent tank

Section	Step	Procedure	Check when finished
D	22	Carefully remove the paper or plate from the tank.	
D	23	If you are using paper, take out the staples and spread the chromatogram out on a paper towel.	
D	24	*Immediately* mark the solvent line with a pencil.	
D	25	Allow the chromatogram to dry completely. A hair dryer or an oven of about 80°C may be used to speed up the drying process.	
D	26	Pour the solvent into a waste solvent container.	
D	27	***Working in the hood,*** spray the paper or plate lightly, but evenly, with a ninhydrin solution. ***Caution: Use the ninhydrin spray inside the hood. Do not breathe the fumes or get spray on your skin.***	
D	28	Dry the sprayed paper or plate by placing it in a drying oven at about 80°C for 3–5 minutes or use a hair dryer. Distinct, colored spots will appear where the ninhydrin reacted with the amino acids.	

Section	Step	Procedure	Check when finished
D.1	1	Draw the chromatogram on the report sheet, or staple the original to the report sheet.	
D.1	2	Record the color of each spot on the drawing or original.	
D.2	1	Measure the distance (cm) from the starting line to the top of the solvent line to obtain the distance traveled by the solvent.	
D.2	2	Record the distance traveled.	

D.3	1	Outline each spot with a pencil.	
D.3	2	Place a dot at the center of each spot.	
D.3	3	Measure the distance in centimeters (cm) from the origin to the center dot of each spot.	
D.4	1	Record the distance.	
D.4	2	Calculate and record the R_f values for the known amino acid samples and the unknown amino acid. $$R_f = \frac{\text{distance traveled by an amino acid}}{\text{distance traveled by the solvent}}$$	
D.5	1	Compare the color and R_f values produced by the unknown amino acids. Identical amino acids will give similar R_f values and form the same color with ninhydrin.	
D.5	2	Identify the amino acid(s) in the unknown.	

Report Sheet - Lab 19

Date _____ Name _____
Section _____ Team _____
Instructor _____ _____

Pre-Lab Study Questions

1. What is the functional group in amines?

2. What are the functional groups in all amino acids?

3. How does an R group determine if an amino acid is acidic, basic, or nonpolar?

Amines and Amino Acids

Report Sheet - Lab 19

Date _____   Name _____
Section _____   Team _____
Instructor _____   _____

A. Solubility of Amines in Water

	Aniline	N-Methylaniline	Triethylamine
A.1 Condensed structural formula			
Classification 1°, 2°, 3°			
A.2 Odor			
A.3 Solubility in water			
A.4 pH			

Questions and Problems

Q.1   What type of compound accounts for the "fishy" odor of fish?

Q.2   Explain why amines are basic.

Laboratory 19

Report Sheet - Lab 19

B. Neutralization of Amines with Acids

	Aniline	N-Methylaniline	Triethylamine
B.1 Solubility after adding HCl			
Odor after adding HCl			

B.2 Equation for the neutralization of aniline with HCl

Equation for the neutralization of N-methylaniline with HCl

Equation for the neutralization of triethylamine with HCl

Questions and Problems

Q.3   How does lemon juice remove the odor of fish?

Q.4   Write an equation for the reaction of butylamine with HCl.

Report Sheet - Lab 19

C. Amino Acids

C.1	Structures of Amino Acids		
	Glycine	Alanine	Serine
		Hydrophobic or hydrophilic?	

C.2 Zwitterion structure of glycine	C.3 Glycine ion in base	Glycine ion in acid

Questions and Problems

Q.5 Write the structure of the zwitterion of alanine.

Laboratory 19

Report Sheet - Lab 19

Q.6 Write the prevalent form of alanine in an acidic solution.

D. Chromatography of Amino Acids

D.1 Draw chromatogram with colors of spots written in or put original paper into the report here.

Calculation of R_f Values

D.2 Distance from origin to solvent line: _____

Amino acid	Color	D.3 Distance (cm) amino acid traveled	D.4 R_f value
Phenylalanine			
Alanine			
Glutamic acid			
Serine			
Lysine			
Aspartic acid			
Unknown			

D.5 Identification of unknown # _____ : _____

20

Peptides and Proteins

Goals
- Identify the structural patterns of proteins.
- Observe the denaturation of proteins.
- Use the isoelectric point of casein in milk to isolate the protein.
- Use chemical tests to identify proteins and amino acids.

Text References

Review the following before performing this laboratory.

Timberlake; *General, Organic, and Biological Chemistry: Structures of Life*, Third Edition
Sections 19.1 to 19.6; Pages 673 to 695

or

Timberlake; *Chemistry: An Introduction to General, Organic, and Biological Chemistry*, Tenth Edition
Sections 16.4 to 16.5; Pages 559 to 570

Discussion

A. Peptide Bonds

A dipeptide forms when two amino acids bond together. A peptide (or amide) bond forms between the carboxylic acid of one amino acid and the amino group of the next amino acid with the loss of H_2O.

$$\underset{\text{Amino acid (1)}}{H_2N-\underset{\underset{H}{|}}{\overset{\overset{R_1}{|}}{C}}-\overset{\overset{O}{\|}}{C}-OH} + \underset{\text{Amino acid (2)}}{H-\underset{\underset{H}{|}}{\overset{\overset{H}{|}}{N}}-\underset{\underset{H}{|}}{\overset{\overset{R_2}{|}}{C}}-\overset{\overset{O}{\|}}{C}-OH} \underset{\text{Hydrolysis}}{\overset{\text{Peptide forms}}{\rightleftarrows}} \underset{\text{Dipeptide}}{H_2N-\underset{\underset{H}{|}}{\overset{\overset{R_1}{|}}{C}}-\overset{\overset{O}{\|}}{C}-\overset{\overset{\text{Peptide bond}}{}}{\underset{\underset{}{}}{N}}-\underset{\underset{H}{|}}{\overset{\overset{R_2}{|}}{C}}-\overset{\overset{O}{\|}}{C}-OH} + H_2O$$

In the reverse reaction, hydrolysis, water adds to the peptide bond to yield the individual amino acids. For example, the dipeptide glycylalanine hydrolyzes to give the amino acids glycine and alanine.

$$\underset{\text{Glycylalanine (Gly-Ala), a dipeptide}}{H_2N-CH-C-N-CH-C-OH} + H_2O \xrightarrow{H^+ \text{ or enzyme}} \underset{\text{Glycine}}{H_2N-CH-C-OH} + \underset{\text{Alanine}}{H_2N-CH-C-OH}$$

B. Structure of Proteins

When many amino acids are joined by peptide bonds they form a polypeptide. If more than 50 amino acids are in the peptide chain, it is usually considered to be a protein. Proteins make up many important features in the body including skin, muscle, cartilage, hair, fingernails, enzymes, and hormones. Proteins have specific structures, which are determined by the sequence of the amino acids. The peptide bonds that join one amino acid to the next are the *primary* level of protein structure. *Secondary* structures include the alpha (α) helix formed by the coiling of the peptide chain, or a β-pleated sheet structure formed between protein strands.

The secondary structures are held in place by many hydrogen bonds between the oxygen atoms of the carbonyl groups and the hydrogen atoms of the amide group. At the *tertiary* level, interactions between the side groups such as ionic bonds or salt bridges, disulfide bonds, and hydrophilic interactions give the protein a compact shape. Such tertiary structures are evident in the spherical shape of globular proteins. Similar interactions between two or more tertiary units produce the *quaternary* structure of many active proteins.

C. Denaturation of Proteins

Denaturation of a protein occurs when certain conditions or agents disrupt the bonds that hold together the secondary or tertiary structures of a protein. Proteins are denatured with heat, acid, base, ethanol, tannic acid, and heavy metal ions (silver, lead, and mercury). In most cases, the protein coagulates. Heat increases the motion of the atoms and disrupts the hydrogen bonds and the hydrophobic (nonpolar) attractions. Strong acids and bases disrupt the salt bridges between amino acids with acidic and basic side groups. Alcohol, an organic solvent, destroys hydrogen bonds. The heavy metal ions, Ag^+, Pb^{2+}, and Hg^{2+}, react with the sulfur groups and carboxylic acid groups, which prevents the formation of the cross-links for tertiary structures and quaternary structures.

D. Isolation of Casein (Milk Protein)

A typical source of protein is milk, which contains the protein casein. When the pH of a sample of nonfat milk is acidified, it reaches its isoelectric point (the same number of positive as negative charges on the molecule), and the protein separates out of the solution. The change in pH disrupts the bonds that hold the tertiary structure together. Adjusting the pH of the mixture causes the casein to solidify so it can be removed. A similar process is used in the making of yogurt, cheeses, and cottage cheese. An enzyme provides the acid for the denaturation of the protein for those products. In this experiment, the mass of a quantity of milk and the mass of isolated casein will be determined. From this data, the percent casein in milk will be calculated.

E. Color Tests for Proteins

Certain tests give color products with amino acids, peptides, and/or proteins. The results of the test can be used to detect certain groups or type of bonds within proteins or amino acids.

Biuret test The biuret test is positive for a peptide or protein with two or more peptide bonds. In the *biuret test*, the blue color of a basic solution of Cu^{2+} turns to a violet color when a tripeptide or larger peptide is present. Individual amino acids and dipeptides do not react with the reagent, and the solution will remain blue (negative).

Ninhydrin test The ninhydrin test is used to detect amino acids and most proteins. In the test, most amino acids produce a blue-violet color. Proline and hydroxyproline give a yellow color.

Xanthoproteic test This test is specific for amino acids that contain an aromatic ring. Concentrated nitric acid reacts with the side chains of tyrosine and tryptophan to give nitro-substituted benzene rings that appear as yellow-colored products.

Lab Information

Time: 2–3 hr

Comments: In protein color tests, be sure to note the color of the reagent before you add it to the samples.

Concentrated HNO_3, 10% Nitric acid, 10% NaOH are extremely corrosive and damage skin and eyes. 10% acetic acid and 10% silver nitrate can stain the skin and clothing.
Tear out the report sheets and place them beside the matching procedures.

Related topics: Amino acids, peptide bonds, structural levels of proteins, denaturation of proteins

Dispose of all chemicals as directed by your laboratory instructor.

Experimental Procedures *GOGGLES REQUIRED!*

A. Peptide Bonds

Materials: Organic model set

Section	Step	Procedure	Check when finished
A.1	1	Make models of glycine and serine.	
A.1	2	Remove the components of water (H—OH) from an amino and carboxylic acid group to form the dipeptides glycylserine and serylglycine.	
A.1	3	Draw each of their structures.	
A.2	1	Demonstrate hydrolysis of the dipeptide serylglycine by breaking the peptide bond and adding the components of H_2O.	
A.2	2	Write a balanced equation for the hydrolysis.	

Laboratory 20

B. Structure of Proteins

Materials: Organic model set

Section	Step	Procedure	Check when finished
B.1	1	Make a model of an amino acid.	
B.1	2	Draw its structure and name it.	
B.1	3	Share your model with two other students, or groups of students who will make models of other amino acids. Line the models up and write their amino acid structures.	
B.2	1	Form peptide bonds between the amino acids by removing the components of water.	
B.2	2	Write an equation for the formation of a tripeptide.	
B.2	3	Use the three letter designation for the amino acids to write the tripeptide order.	
B.2	4	Rearrange the amino acids and prepare a different tripeptide.	
B.2	5	Write the structure and name of the new tripeptide.	

C. Denaturation of Proteins

Materials: Test tubes, test tube holder, 10-mL graduated cylinder, 1% egg albumin (or an egg, cheesecloth, and beaker), 10% HNO_3, 10% NaOH, 95% ethyl alcohol, 1% $AgNO_3$ (dropper bottle)

Section	Step	Procedure	Check when finished
C	1	Mix the white from one egg with 200 mL of water or use the 1% egg albumin solution.	
C	2	If making the egg albumin solution fresh, filter the mixture through cheesecloth into a beaker.	
C	3	Place 2–3 mL of egg albumin solution in each of five test tubes. Use one sample for each of the following tests.	

Record your observations and give a brief explanation for the results.

Section	Step	Procedure	Check when finished
C.1	1	**Heat** Using a test tube holder, in a different test tube, heat the egg albumin solution over a low flame.	
C.1	2	Record the changes, if any.	

Peptides and Proteins

Section	Step	Procedure	Check when finished
C.2	3	**Acid:** Add 2 mL of 10% HNO_3.	
C.2	4	Record the changes, if any.	
C.3	1	**Base:** Add 2 mL of 10% NaOH.	
C.3	2	Record the changes, if any.	
C.4	1	**Alcohol:** Add 4 mL of 95% ethyl alcohol. Mix.	
C.4	2	Record the changes, if any.	
C.5	1	**Heavy metal ions:** Add 10 drops of 1% $AgNO_3$.	
C.5	2	Record the changes, if any.	
C.6	1	Lab detergent: Add 10 drops of lab soap or solution and stir.	
C.6	2	Record the changes, if any.	

D. Isolation of Casein (Milk Protein)

Materials: 150-mL beaker, hot plate or Bunsen burner, thermometer, funnel or Büchner filtration apparatus, filter paper, watch glass, nonfat milk, 10% acetic acid, dropper, pH paper, stirring rod

Section	Step	Procedure	Check when finished
D.1	1	Weigh a 150-mL beaker.	
D.1	2	Record the weight.	
D.1a	1	Add about 20 mL of nonfat milk to the beaker.	
D.1a	2	Weigh the beaker and nonfat milk.	
D.1b	1	Record the weight.	
D.1b	2	Calculate the mass of the nonfat milk sample (D.1a – D.1)	
D.2	1	Using pH indicator paper, determine the pH of the milk sample.	

Section	Step	Procedure	Check when finished
D.3	1	Warm the sample on a hot plate or a Bunsen burner until the temperature reaches about 50°C.	
D.3	2	Remove the beaker and milk from the heat.	
D.3	3	Add 10% acetic acid, drop by drop. You may need 2 to 3 mL. Using a stirring rod, stir continuously. At the isoelectric point, the casein (milk protein) becomes insoluble. When no further precipitation occurs, stop adding acid.	
D.3	4	If the liquid layer is not clear, heat the mixture gently for a few more minutes.	

Laboratory 20

Section	Step	Procedure	Check when finished
D.3	5	Using pH indicator paper, determine the pH at which the casein becomes insoluble in solution. This is the pH of the isoelectric point of casein.	
D.3	6	Record the pH.	

Section	Step	Procedure	Check when finished
D.4	1	Collect the solid protein using a funnel and filter paper or the Büchner filtration apparatus.	
D.4	2	Wash the protein with two 10-mL portions of water.	
D.4	3	Weigh a watch glass.	
D.4	4	Record the weight.	
D.4a	1	Transfer the protein to the watch glass and let the protein dry.	
D.4a	2	Weigh the protein and the watch glass.	
D.4a	3	Record the weight.	
D.4b	1	Calculate the mass of casein (D.4a – D.4) Save the casein for part E.	
D.5	1	Calculate the percentage of casein in the nonfat milk. $$\% \text{ Casein} = \frac{\text{mass (g) of casein}}{\text{mass (g) of milk}} \times 100\%$$	

E. Color Tests for Proteins

Materials: Test tubes, test tube rack, 10-mL graduated cylinder, boiling water bath, cold water bath, pH paper, spatula, dropper bottles of 1% amino acid solution (glycine, tyrosine), dropper bottles of 1% solutions of proteins (gelatin, egg albumin), casein from part D, 0.2% ninhydrin solution, concentrated HNO_3 (dropper bottle), 10% NaOH, 5% $CuSO_4$ (biuret), red litmus paper

Section	Step	Procedure	Check when finished			
E.1	1	**Biuret test:** Place 2 mL of the solution of glycine in a test tube.				
E.1	2	Add 2 mL of 10% NaOH.				
E.1	3	Stir.				
E.1	4	Add 5 drops of biuret reagent (5% $CuSO_4$).				
E.1	5	Stir.				
E.1	6	Record the color.				

Peptides and Proteins

Section	Step	Procedure	Check when finished
E.1	7	The formation of a pink-violet color indicates the presence of a protein with two or more peptide bonds. If such a protein is not present, the blue color of the cupric sulfate will remain (negative result).	
E.1	8	Record the results.	
E.1	9	Repeat steps 1 to 8 substituting the amino acid tyrosine, gelatin, egg albumin and a small amount of solid casein (from part D), the amount held on the tip of a spatula.	
E.1	10	Record the results.	

Section	Step	Procedure	Check when finished
E.2	1	**Ninhydrin test:** Place 2 mL of the solution of glycine in a test tube.	
E.2	2	Add 1 mL of 0.2% ninhydrin solution.	
E.2	3	Place the test tubes in a boiling water bath for 4–5 minutes.	
E.2	4	Record your observations. Look for the formation of a blue-violet color.	
E.2	5	Repeat steps 1 to 4 substituting the amino acid tyrosine, gelatin, egg albumin and a small amount of solid casein (from part D), the amount held on the tip of a spatula.	
E.2	6	Record the results.	

Section	Step	Procedure	Check when finished
E.3	1	**Xanthoproteic test** *(optional)*: Place 2 mL of the solution of glycine in a test tube.	
E.3	2	*Cautiously* add 10 drops of concentrated HNO_3 to the sample.	
E.3	3	Place the test tubes in a boiling water bath and heat for 3–4 minutes.	
E.3	4	Remove the test tubes, place them in cold water, and let them cool.	
E.3	5	Carefully add 10% NaOH, drop by drop, until the solution is just basic (turns red litmus blue). This may required 2–3 mL of NaOH. *Caution: heat will be evolved.*	
E.3	6	Record your observations. Look for the formation of a yellow-orange color, which may vary in intensity.	

E.3	7	Repeat steps 1 to 4 substituting the amino acid tyrosine, gelatin, egg albumin and a small amount of solid casein (from part D), the amount held on the tip of a spatula.				
E.3	8	Record your results.				

Report Sheet - Lab 20

Date _____ Name _____
Section _____ Team _____
Instructor _____   _____

Pre-Lab Study Questions

1. What is a peptide bond?

2. How does the primary structure of proteins differ from the secondary structure?

Laboratory 20

Report Sheet - Lab 20

Date _____ Name _____
Section _____ Team _____
Instructor _____

A. Peptide Bonds

A.1   Structure of glycylserine

       Structure of serylglycine

A.2   Hydrolysis of serylglycine

B. Structure of Proteins

B.1   Amino acid structures and names

B.2   Equation for the formation of the first tripeptide

Laboratory 20

Report Sheet - Lab 20

Order of amino acids using three letter code

Equation for the formation of the second tripeptide

Order of amino acids using symbols

C. Denaturation of Proteins

Treatment	Observations of Egg Albumin	Explanation
C.1 Heat		
C.2 Acid		
C.3 Base		
C.4 Alcohol		
C.5 Heavy metal ions		
C.6 Lab detergent		

Questions and Problems

Q.1 Why are heat and alcohol used to disinfect medical equipment?

Report Sheet - Lab 20

Q.2   Why is milk given to someone who accidentally ingests a heavy metal ion such as silver or mercury?

D.   Isolation of Casein (Milk Protein)

D.1	Mass of beaker	
D.1a	Mass of beaker and milk	
D.1b	Mass of milk	
D.2	pH of milk	
D.3	pH when casein precipitates	
D.4	Mass of watch glass	
D.4a	Mass of watch glass and casein	
D.4b	Mass of casein	
D.5	Percent casein *(Show calculations.)*	

Questions and Problems

Q.3   Compare the pH of the milk sample and the pH at which the casein solid forms.

Q.4   How does a change in pH affect the structural levels of a protein?

Laboratory 20

Report Sheet - Lab 20

E. Color Tests for Proteins

Observations of Color Tests			
Sample	E.1 Biuret	E.2 Ninhydrin	E.3 Xanthoproteic
Glycine			
Tyrosine			
Gelatin			
Egg albumin			
Casein (milk protein)			

Questions and Problems

Q.5   After working with HNO_3, a student noticed that she had a yellow spot on her hand. What might be the reason?

Q.6   Which samples give a negative biuret test? Why?

Q.7   What functional group gives a positive test in the xanthoproteic test?

Q.8   What tests could you use to determine whether an unlabeled test tube contained an amino acid or a protein?

21

Enzymes

Goals

- Prepare a solution of the enzyme amylase.
- Describe the role of an enzyme as a catalyst in biological systems.
- Set up chemical tests that indicate the rate of an enzyme-catalyzed reaction.
- Observe the effects of enzyme concentration, temperature, pH, and inhibitors upon enzyme activity.

Text References

Review the following before performing this laboratory.

Timberlake; *General, Organic, and Biological Chemistry: Structures of Life*, Third Edition
Sections 20.1 to 20.5; Pages 703 to 721

or

Timberlake; *Chemistry: An Introduction to General, Organic, and Biological Chemistry*, Tenth Edition
Sections 16.6 to 16.8; Pages 571 to 580

Discussion

In biological systems, reactions are catalyzed by enzymes, which speed up reactions while operating at mild conditions of temperature and pH. Like all catalysts, enzymes lower the energy of activation needed for a reaction to take place. For example, an enzyme in blood called carbonic anhydrase converts carbon dioxide and water to carbonic acid. The enzyme catalyzes the reaction of about 35 million molecules of CO_2 every minute. See Figure 21.1.

Figure 21.1 Activation energy of a reaction with and without an enzyme (catalyst)

In this experiment you will use amylase, an enzyme that begins the hydrolysis of carbohydrates (amylose) in the mouth. In the presence of amylase, a sample of starch will undergo hydrolysis to give smaller polysaccharides, dextrins, maltose, and glucose.

Laboratory 21

$$\text{Starch (amylose)} \xrightarrow{\textit{Amylase}} \text{Smaller polysaccharides, dextrins, maltose} \xrightarrow{\textit{Amylase}} \text{Glucose}$$

Testing Enzyme Activity

We will determine the reaction of amylase with starch by testing for starch using iodine reagent. When iodine is added to a starch solution, a blue-black color is produced. However, if amylase is present and the starch is hydrolyzed, the iodine test is no longer positive. Only the red or gold color of the iodine solution is seen. The faster the amylase breaks down starch, the more quickly the blue-black color is lost. If the blue-black color persists, the enzyme is inactive.

The hydrolyzed product glucose can be detected by using Benedict's reagent. With glucose, the Benedict's reagent turns from blue to a green or reddish-orange with higher glucose concentrations. Table 21.1 summarizes the tests for amylose and glucose.

Table 21.1 *Detection Tests for Starch and Glucose*

Starch (Amylose)	Glucose
Positive iodine test (Turns deep blue with iodine)	Negative iodine test (No color change with iodine. Remains red or gold)
Negative Benedict's test (Remains blue)	Positive Benedict's test (Turns green to reddish-orange)

A. Effect of Enzyme Concentration

During catalysis, an enzyme combines with the reactant or *substrate* of a reaction to give an *enzyme–substrate complex*. To form this complex, the substrate fits into the *active site,* where reaction takes place. The *products* are released and the enzyme is ready to catalyze another reaction.

$$\underset{\text{Enzyme}}{E} + \underset{\text{Substrate}}{S} \rightleftarrows \underset{\text{Enzyme–substrate complex}}{ES} \rightleftarrows \underset{\text{Enzyme}}{E} + \underset{\text{Product}}{P}$$

If the enzyme concentration is increased while substrate concentration is constant, the rate of the reaction will increase. With more enzyme present, more substrate molecules can react.

B. Effect of Temperature

The optimum temperature is the temperature at which an enzyme operates at maximum efficiency. At low temperatures, the rate of reaction is slowed. At high temperatures, the enzyme protein is denatured. See Figure 21.2.

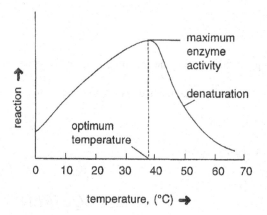

Figure 21.2 Effect of temperature on enzyme activity

C. Effect of pH

At the optimum pH, an enzyme is most active. At pH values above and below optimum, the protein structure of the enzyme is altered, which can severely reduce the enzyme's activity. See Figure 21.3.

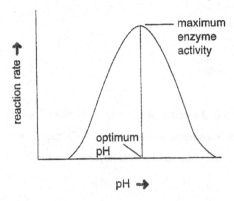

Figure 21.3 Effect of pH on enzyme activity

D. Inhibition of Enzyme Activity

Substances that limit or stop the catalyzing activity of an enzyme are called *inhibitors*. A *competitive inhibitor* competes for the active site of an enzyme, while a *noncompetitive inhibitor* binds to the surface of the enzyme at another site and disrupts the structure of the active site. An *irreversible inhibitor* forms bonds with side-chains of the amino acids in the active site, which makes the enzyme inactive while a *reversible inhibitor* can disassociate from the enzyme and restore enzyme activity.

Laboratory 21

Lab Information

Time: 2½ hr

Comments: Tear out the report sheets and place them beside the matching procedures.

Related topics: Enzymes, active site, lock-and-key theory, enzyme activity, factors affecting enzyme activity, inhibition, denaturation

Dispose of all chemicals as directed by your laboratory instructor.

Experimental Procedures   *GOGGLES REQUIRED!*

Constant temperature water baths: If commercial water baths are not available, prepare your own using a hot plate. Fill three 250-mL or 400-mL beakers about half full of tap water. Warm one to about 37°C, and heat one to boiling. To the third, add ice to give an ice bath with a low temperature (0 to 5°C)

Type of Bath	Temperature of Water Baths
Boiling hot water bath	100°C
Warm water	37°C (body-temperature)
Ice bath	0° (or lower than 5°C)

Starch solution: Place 50 mL of 1% starch solution in a small beaker.

Iodine reagent: Obtain a small amount of iodine reagent in a second small beaker. Dropper bottles of iodine may be available. In a third small beaker, place 40 mL of a freshly prepared amylase solution. Obtain a spot plate or plastic sheet for testing. Use clean droppers and rinse. Rinsing the droppers is important to avoid transferring enzyme to other samples.

Amylase preparation: Your instructor may provide a commercial amylase solution.

Reference Tests
Starch: Place a few drops of starch solution in a depression in the spot plate. Add 1 drop of iodine reagent. The reaction with starch should give a deep blue-black color.

Glucose sample: Place 3 mL of 1% glucose in a test tube. Add 2 mL of Benedict's solution and heat for 3–4 minutes in a boiling water bath. The reaction should give a solid with a red-orange color.

Visual Color Reference

As you proceed with each experiment, you will check enzyme activity by adding iodine to the starch mixture with iodine. When enzyme activity is high, the time required for the starch to hydrolyze will be very short. When the enzyme is slowed down or inactive, the blue-black color will be seen for a longer time. By

Enzymes

observing the disappearance of starch, you can assess the relative amount of enzyme activity as using Table 21.2.

Table 21.2 *Iodine test color and enzyme activity comparison table.*

Iodine Test for Starch	Amount of Starch Remaining	Enzyme	Activity Level
Dark blue-black	All	None	0
Blue	Most	Low	1
Light brown	Some	Moderate	2
Gold	None	High	3

A. Effect of Enzyme Concentration

Materials: Test tubes, test tube rack, thermometer, 37°C water bath, droppers, spot plate (or plastic sheet), amylase preparation, 1% starch (buffered pH 7.0), iodine reagent, Benedict's solution, boiling water bath, 5- or 10-mL graduated cylinder

Section	Step	Procedure	Check when finished
A.1	1	Fill a beaker which can hold 4 test tubes, about 2/3 full of tap water.	
A.1	2	Warm to about 37°C and try to maintain the temperature. Your lab may have a commercial water bath set at 37°C.	
A.1	3	Place 4 mL of 1% starch in each of four separate test tubes labeled 1–4.	
A.1	4	Place 4 mL of amylase solution in a fifth tube.	
A.1	5	Place all the tubes in a 37°C water bath for 5 minutes.	
A.1	6	Tube 1 without enzyme is the control. Do not put anything in that tube.	
A.1	7	Add 3 drops of the warmed amylase solution to Tube 2.	
A.1	8	Add 6 drops of the warmed amylase solution to Tube 3.	
A.1	9	Add 9 drops of the warmed amylase solution to Tube 4.	
A.1	10	Mix quickly by shaking the test tube.	
A.1	11	Return the test tubes immediately to the 37°C water bath.	
A.1	12	Record the time at which you add enzyme.	
A.1	13	**Immediately**, transfer four drops of each reaction mixture (use clean droppers) to a spot plate (or plastic sheet).	
A.1	14	Add 1 drop of iodine reagent to each.	

Laboratory 21

Section	Step	Procedure	Check when finished		
A.1	15	Record your observations for each sample. Use the visual color reference to assess the enzyme activity.			
A.1	16	Rinse the spot plate or use a new section of plastic sheet.			
A.1	17	**Repeat the test again in 5 and 10 minutes.**			

(Optional)

Section	Step	Procedure	Check when finished
A.2	1	Add 3 mL of Benedict's solution to each test tube and place them in a boiling water bath for 3–4 minutes.	
A.2	2	Record the color. The appearance of a green to orange-rust color indicates the presence of glucose. If the solution remains clear blue, no hydrolysis of the starch has taken place.	

Section	Step	Procedure	Check when finished
A.3	1	Make a graph comparing the enzyme activity level using the relative scale of enzyme activity given in Table 32.1 and the amount of amylase solution at 10 minutes.	

B. Effect of Temperature

Materials: Test tubes, test tube rack, test tube holder, droppers, 5- or 10-mL graduated cylinder, beakers for water baths: ice (0°C), warm (37°C), and boiling (100°C), amylase preparation, spot plate (or plastic sheet), 1% starch, iodine reagent

Section	Step	Procedure	Check when finished		
B.1	1	Place 4 mL of 1% starch solution in each of three test tubes.			
B.1	2	Place one tube in a boiling water bath.			
B.1	3	Place one tube in the 37°C water bath.			
B.1	4	Place one tube in the ice bath.			
B.1	5	Add 4 mL of amylase solution to three other test tubes and place one each in the three water baths. Let them remain in the water baths for about 10 minutes to allow the solutions to reach the bath temperature.			
B.1	6	For the boiling water bath, record the temperature of the bath.			
B.1	7	Remove the test tubes, pour the contents together in one tube.			
B.1	8	Mix by shaking.			

Enzymes

Section	Step	Procedure	Check when finished
B.1	9	Return the mixture to the same temperature bath.	
B.1	10	Repeat steps 6 through 9 for the 37°C water bath and the ice bath.	
B.1	11	Wait 10 minutes.	
B.1	12	Transfer four drops of the mixture from the boiling water bath to a spot plate (or plastic sheet).	
B.1	13	Add 1 drop of iodine.	
B.1	14	Record the color and activity level.	
B.1	15	Repeat steps 12 through 14 for the 37°C water bath and the ice bath.	
B.2	1	Make a graph of the enzyme activity level using the relative scale of enzyme activity given in Table 32.1 and the temperature.	

C. Effect of pH

Materials: Test tubes, test tube rack, amylase preparation, 37°C water bath, buffers (pH 2, 4, 7, 10), spot plate (or plastic sheet), 1% starch, iodine reagent

Section	Step	Procedure	Check when finished
C.1	1	Place 4 mL of buffer solutions of pH 2, 4, 7, and 10 in separate test tubes.	
C.1	2	Add 4 mL of amylase solution to each.	
C.1	3	In four other test tubes place 4 mL of 1% starch solution.	
C.1	4	Place all of the test tubes in a 37°C water bath for about 5 minutes.	
C.1	5	Pour each of the 1% starch solutions into a separate buffer–amylase tube.	
C.1	6	Mix by shaking and return the mixtures to the 37°C water bath.	
C.1	7	Wait 15 minutes.	
C.1	8	Remove 4 drops from the pH 2 test tube and place in a spot plate or on the plastic sheet.	
C.1	9	Add 1 drop of iodine reagent.	
C.1	10	Record your observations and the enzyme activity level.	
C.1	11	Repeat steps 8 through 10 for the pH 4, 7 and 10 test tubes.	
C.2	1	Make a graph of the enzyme activity level using the relative scale of enzyme activity given in Table 32.2 and pH.	

Laboratory 21

D. Inhibition of Enzyme Activity

Materials: Test tubes (6), test tube rack, 5- or 10-mL graduated cylinder, amylase preparation, 37°C water bath, solutions in dropper bottles: 1% NaCl, 1% $AgNO_3$, 1% $Pb(NO_3)_2$, 1% $HgCl_2$, 95% ethanol, 1% starch, iodine reagent, spot plate (or plastic sheet)

Section	Step	Procedure	Check when finished
D.1	1	In one test tube, place 4 mL of amylase solution and 10 drops of 1% NaCl.	
D.1	2	In a second test tube, place 4 mL of amylase solution and 10 drops of 95% ethanol.	
D.1	3	In a third test tube, place 4 mL of amylase solution and 10 drops of one of the inhibitors that your instructor assigns: 1% $AgNO_3$, 1% $Pb(NO_3)_2$, or 1% $HgCl_2$.	
D.1	4	Record the identity of the inhibitor.	
D.1	5	Place 4 mL of 1% starch in three other test tubes.	
D.1	6	Place all six tubes in a 37°C water bath.	
D.1	7	Wait 5 minutes.	
D.1	8	Combine the starch solutions with the amylase solutions, mix, and return the tubes to the 37°C bath.	
D.1	9	Wait 15 minutes.	
D.1	10	Remove four drops of the reaction mixture with NaCl (use clean droppers each time) and place in the spot plate or on the plastic sheet.	
D.1	11	Add 1 drop of iodine reagent.	
D.1	12	Record your observations and the enzyme activity level.	
D.1	13	Repeat steps 10 through 12 for the test tubes with ethanol and inhibitor.	

Enzymes

Report Sheet - Lab 21

Date _____ Name _____
Section _____ Team _____
Instructor _____   _____

Pre-Lab Study Questions

1. What is the substrate of the enzyme amylase?

2. What are the products of amylase action?

3. What happens to enzymes at high temperatures?

4. Reference tests:

   Color of iodine test for starch   _____

   Color of Benedict's test for glucose   _____

Enzymes

Report Sheet - Lab 21

Date _____ Name _____
Section _____ Team _____
Instructor _____ _____

A. Effect of Enzyme Concentration

A.1   Time when you put the enzyme in the first test tube _____

A.1 Time	Color Produced by Iodine Test and Activity Level			
	No amylase	3 drops amylase	6 drops amylase	9 drops amylase
0 minutes				
5 minutes				
10 minutes				

	Color Produced by Benedict's Test			
A.2 Benedict's test				

A.3   Graph: Enzyme activity at 10 minutes vs. enzyme concentration

Drops of amylase

Laboratory 21

Report Sheet - Lab 21

Questions and Problems

Q.1   At which enzyme concentration was starch hydrolyzed the fastest? Slowest?

Q.2   Describe the effect of enzyme concentration on enzyme activity.

B.   Effect of Temperature

B.1

Color Produced by Iodine Test and Activity Level					
0°C	Level	37°C	Level	100°C	Level

B.2   Graph: Enzyme activity level and temperature

Enzymes

Report Sheet - Lab 21

Questions and Problems

Q.3   What was the optimal temperature for amylase?

Q.4   Why did the enzyme activity differ at 0°C and at 100°C?

C.   Effect of pH

C.1   Color Produced by Iodine Test and Activity Level

	Color Produced by Iodine Test and Activity Level							
	pH 2	Level	pH 4	Level	pH 7	Level	pH 10	Level

C.2   Graph: Enzyme activity level and pH

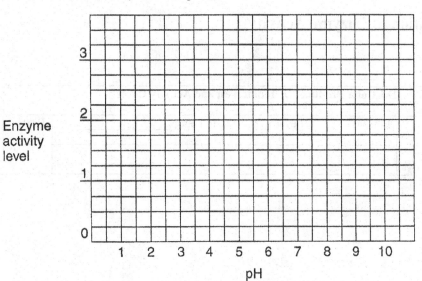

Laboratory 21

Report Sheet - Lab 21

Questions and Problems

Q.5 How is amylase affected by a low pH? By a high pH? Explain.

Q.6 What was the optimum pH for amylase?

Q.7a. During digestion, the pH in the stomach is 2. What does this indicate about the optimum pH of pepsin, an enzyme that hydrolyzes protein in the stomach?

Q.7b. What happens to the activity of pepsin when it enters the small intestine where the pH is 8?

D. Inhibition of Enzyme Activity

D.1 Color Produced by Iodine Test and Activity Level					
NaCl	Level	Ethanol	Level	Inhibitor _____	Level

Report Sheet - Lab 21

Questions and Problems

Q.8 In which reaction mixture(s) did hydrolysis of starch occur?

Q.9 What substances added to the mixture were inhibitors?

Q.10 How might those substances inhibit enzyme action?

Q.11 What are some differences and/or similarities in the type of inhibition caused by heat, acid or base, and heavy metal ions on enzyme activity?

22

DNA Components and Extraction

Goals

- Identify the components in DNA: purines and pyrimidines, polyphosphate, and deoxyribose.
- Identify the monomer units in DNA.
- Extract and compare DNA samples from a variety of plant and animal cells.

Text References

Review the following before performing this laboratory.

Timberlake; *General, Organic, and Biological Chemistry: Structures of Life*, Third Edition
Sections 21.1 to 21.3; Pages 739 to 747

or

Timberlake; *Chemistry: An Introduction to General, Organic, and Biological Chemistry*, Tenth Edition
Sections 17.1 to 17.3; Pages 592 to 601

Discussion

In the cells of all living things are molecules called nucleic acids that provide the information for cellular replication and growth by directing protein synthesis. There are two main types of nucleic acids: DNA (deoxyribonucleic acid), which carries the genetic code to each new generation of cells, and RNA (ribonucleic acid), which carries the instructions for a protein to the ribosomes where proteins are synthesized.

Every living organism contains DNA that provides the directions to make the proteins for the characteristics of that organism. The DNA in an onion has a set of directions to make an onion, whereas the DNA in a banana directs the growth of a banana. Each daughter cell is just like the parent cell because the DNA is duplicated in every cell division.

A. Components of DNA

Nucleic acids are polymers of repeating units known as nucleotides. There are millions of nucleotides in a single DNA molecule. RNA molecules are smaller and contain several thousand nucleotides. A nucleotide of DNA consists of a 5-carbon deoxyribose sugar, one of four nitrogenous bases, and a phosphate group. (See Figure 22.1.)

Laboratory 22

Deoxyadenosine 5'–monophosphate (dAMP)

Figure 22.1 One of four nucleotides found in DNA

In DNA, two bases are the purines adenine (A) and guanine (G) and the other two bases are the pyrimidines cytosine (C) and thymine (T). The nucleotides in RNA differ slightly; the sugar is ribose, and the pyrimidine uracil replaces thymine.

Pyrimidines: Cytosine (C) (DNA and RNA), Thymine (T) (DNA) only, Uracil (U) (RNA only)

Purines: Adenine (A) (DNA and RNA), Guanine (G) (DNA and RNA)

B. Extraction of DNA

In plant cells, DNA strands are combined with protein and RNA molecules. The extraction and isolation of DNA from a cell requires three processes: (1) breaking down the cellular membranes, (2) heating the mixture and denaturing the proteins, and (3) precipitating the DNA as a white, stringy, fibrous material.

Step 1 Breaking down cell membranes

The plant or animal material containing the DNA is mixed with an extraction buffer and homogenized to break down the cell membranes. The extraction buffer consists of sodium dodecyl sulfate (SDS), ethylenediamine tetraacetic acid (EDTA), sodium chloride (NaCl), and sodium citrate. SDS disrupts the polar attractions that hold the cell membrane together. EDTA, a chelating agent, removes Mg^{2+} and Ca^{2+} ions that are needed by nucleases that degrade DNA. The Na^+ binds with the negatively charged phosphate groups in the DNA fragments, which will cause them to precipitate out of an ethanol solution in step 3. The buffer and the following heat treatment cause lipids and proteins to precipitate out of solution.

DNA Components and Extraction

Step 2 Heating and denaturing proteins

Heating denatures proteins and inactivates the nuclease enzymes that degrade DNA. If the nucleases remain active, they would hydrolyze DNA into its nucleotide components. When we cook, we heat foods to denature the protein. In addition, some people add a meat tenderizer, which contains a protease, to break down proteins further and make them more digestible.

Step 3: Precipitating DNA from alcohol solution

The DNA is obtained by adding a cold ethanol solution that causes the DNA fragments to come together and precipitate. Because the DNA threads are sticky, they bind to a glass rod so they can be removed from the solution. Following the extraction of DNA, students will make observations and discuss the similarities and differences in DNA samples obtained from different sources.

Lab Information

Time: 2 to 3 hr
Comments: Tear out the report sheets and place them beside the matching procedures.
Related topics: Deoxyribonucleic acid, nitrogenous bases, deoxyribose, nucleosides, nucleotides, replication

Dispose of all chemicals as directed by your laboratory instructor.

Experimental Procedures   *GOGGLES REQUIRED!*

A.   Components of DNA

Materials: Organic model kits

Section	Step	Procedure	Check when finished
A.1	1	**Making a model of a purine:** Use the C, H, N, and O atoms in a model kit to make a model of one of the purines in DNA nucleotides.	
A.1	2	Have another lab team make a model of the other purine.	
A.1	3	Draw the structures of the purines found in the nucleotides of DNA.	
A.1	4	Save these models.	

Laboratory 22

Section	Step	Procedure	Check when finished
A.2	1	**Making a model of deoxyribose:** Use the C, H, and O atoms in a model kit to make a model of deoxyribose.	
A.2	2	Draw the structure of deoxyribose.	
A.2	3	Save this model.	

Section	Step	Procedure	Check when finished
A.3	1	**Making a model of a phosphate:** Use the P, H and O atoms in a model kit to make a model of phosphate.	
A.3	2	Attach the phosphate group and the purine you constructed to the deoxyribose sugar.	
A.3	3	Draw the structure of this nucleotide.	
A.3	4	Write its name.	
A.3	5	Save this model.	

Section	Step	Procedure	Check when finished
A.4	1	**Making a dinucleotide:** With your neighbor lab team, combine two nucleotides to form a dinucleotide.	
A.4	2	Draw the structure of this dinucleotide.	
A.4	3	Write its name.	
A.4	4	Combined the two nucleotides to make a different dinucleotide.	
A.4	5	Draw the structure of this dinucleotide.	

B. Extraction of DNA

Different student teams may extract DNA from different DNA sources and compare results or the procedure may be run initially with onions and then repeated using a DNA source selected by each student team.

    Materials: **DNA sources:** white onions, variety of other DNA plant sources such as cauliflower, broccoli, garlic, split peas, bananas, and/or animal sources such as chicken liver, calf thymus

    Lab items: Knife, two 250-mL beakers, thermometer, hot water bath (400-mL beaker about $^1/_2$ full of water, iron ring, wire screen, and Bunsen burner), blender, cheesecloth or filter paper, microscope, stirring rod

    Lab chemicals: SDS-NaCl-EDTA buffer solution (Your instructor will prepare this buffer: 50 g sodium dodecyl sulfate, 50 g NaCl, 5 g sodium citrate, 0.2 mL of 0.5 MEDTA and

DNA Components and Extraction

water to make 1 L), isopropanol (placed in an ice bath to keep cold), citrate buffer (0.15 M NaCl, 0.015 M sodium citrate)

Section	Step	Procedure	Check when finished
B	1	**Preparing DNA:** Obtain about 50 g of a white onion or other DNA source.	
B	2	Use a knife to dice the onion or other DNA source into small pieces.	
B	3	Place the pieces in a 250-mL beaker.	
B	4	Add 50 mL of the extraction (SDS-NaCl-EDTA) buffer.	
B	5	Prepare a hot water bath and heat it to 60°C	
B	6	Place the beaker in the hot water bath.	
B	7	Allow the beaker to remain in water at 60°C for 15 minutes. (Any longer time in hot water will start to break down DNA.)	
B	8	Stir the mixture occasionally with a stirring rod.	
B	9	Remove the beaker and place it in an ice bath for 10 minutes.	
B	10	Pour the cooled mixture into a blender and blend the contents for 60 seconds using 15-second bursts.	
B	11	Pour the mixture through two pieces of cheesecloth placed over a clean 250-mL beaker. This filtering may take as long as one hour. You may need to use new sets of cheesecloth as it becomes clogged with cell debris.	
B	12	Measure the volume of the filtered onion liquid collected in the beaker.	
B	13	Obtain an equal volume of ice-cold isopropanol. If the alcohol is not ice cold, cool it first in an ice bath.	
B	14	Holding the beaker at an angle, slowly pour the alcohol down the side.	
B	15	Allow the solution to sit for 2 minutes. An alcohol layer should form on the top of the filtrate. Because DNA is not soluble in alcohol, the whitish, viscous strands of DNA should precipitate out of the alcohol layer and form a whitish interface. (Other components of the mixture will remain soluble in the alcohol layer.)	
B	16	Place a glass rod into the solution and turn it slowly to wind the DNA threads into a ball on the end of the glass rod. When you remove the glass rod the DNA will look like a viscous blob.	

Laboratory 22

B	17	Transfer the DNA from the glass rod to a paper towel and let it dry.	
B	18	Observe the DNA obtained by other student teams with different DNA sources, or repeat the procedure above with another DNA source.	

Section	Step	Procedure	Check when finished
B.1	1	**Observing DNA:** Observe texture and physical properties of the DNA you have extracted.	
B.1	2	Record your observations.	
B.2	1	Examine the DNA under a microscope. Your laboratory instructor will provide instruction on how to do this.	
B.2	2	Record your observations.	

DNA Components and Extraction

Report Sheet - Lab 22

Date _____ Name _____
Section _____ Team _____
Instructor _____   _____

Pre-Lab Study Questions

1. Why does DNA have a slight negative charge?

2. What would homogenization and heating do to cell membranes?

DNA Components and Extraction

Report Sheet - Lab 22

Date _____ Name _____
Section _____ Team _____
Instructor _____   _____

A. Components of DNA

Structures of Models (Lab Team 1)	Structures of Models (Lab Team 2)
A.1	A.1
A.2	A.2
A.3	A.3
A.4	

Laboratory 22

Report Sheet - Lab 22

B. Extraction of DNA

DNA Source B.1	Texture and Appearance	B.2 Appearance under a microscope

Questions and Problems

Q.1 Why is it necessary to heat the DNA source and buffer mixture?

Report Sheet - Lab 22

Q.2 Why was an alcohol added to the filtered onion liquid?

Q.3 What similarities did you observe in the appearance of DNA from different DNA sources? Why?

Q.4 How would be DNA extracted from an onion be different from the DNA extracted from cauliflower? Why?

Q.5 Write the complementary base sequence for the following segment of DNA.

Laboratory 22

Appendix
Materials and Solutions

Standard Laboratory Materials

The equipment and chemicals listed throughout the appendix are the materials needed in the laboratory to perform the experiments in this lab manual. The amounts given are recommended for 20–24 students working in teams. The following equipment is expected to be in the laboratory lockers or available from the laboratory stock and will not be listed in each experiment.

Aspirators
Beakers (50–400 mL)
Balances (top loading or centigram)
Büchner filtration apparatus and filter paper
Bunsen burners
Burets (50 mL)
Buret clamps
Clay triangles
Containers for waste disposal
Crucible and cover
Distilled water (special faucet or containers)
Droppers
Evaporating dish
Flask, Erlenmeyer (125–250 mL)
Filter paper for funnels
Funnel
Glass stirring rods
Gloves
Graduated cylinders (5–250 mL)

Hot plates
Ice
Iron rings
Litmus paper
Meter sticks
pH paper
Ring stand
Rulers
Shell vials
Stirring rod, glass
Spatulas
Stoppers
Test tubes (6", 8")
Test tube rack
Thermometer
Tongs (crucible and beaker)
Watch glass
Wire gauze

Additional Materials Needed for Individual Experiments

The equipment and chemicals listed in this section must be supplied by the instructor or stockroom.. The amounts given are recommended for 20–24 students working in teams of two students

Dry Labs Supplies

D1 Conversion Factors in Calculations
    A. Rounding Off
        10   Solid objects with regular shapes
    D. Conversion Factors for Volume
        3   1-liter graduated cylinders
        3   1-quart (or two 1-pint) measures
    E. Conversion Factors for Mass
        5   Commercial products with mass given on label in metric and U.S. system units
    G. Converting Temperature
        200 g Rock salt

D2 Compounds and their Formulas
    Merck Index or *CRC Handbook of Chemistry and Physics*
    B. Ionic Compounds and Formulas
        $NaCl(s)$

Appendix

    C. **Ionic Compounds with Transition Metals**
        $FeCl_3(s)$
    D. **Ionic Compounds with Polyatomic Ions**
        $K_2CO_3(s)$
    E. **Covalent (Molecular) Compounds**
        H_2O

D3 Properties of Organic Compounds
    A. **Color, Odor, and Physical State** *(May be a display in lab)*
        1–2   Chemistry handbook   20 g $NaCl(s)$
        20 g $KI(s)$ 20 g   Benzoic acid(s)
        20 mL   Toluene   20 mL   cyclohexane
    B. **Solubility** *(This may be an instructor demonstration.)*
        20 g   NaCl(s)   20 mL   Toluene
        20 mL   Cyclohexane
    C. **Combustion** *(This may be an instructor demonstration.)*
        30   Wood splints   10 g $NaCl(s)$
        10 mL   Cyclohexane
    D. **Structures of Alkanes, E. Isomers, F. Cycloalkanes**
        10   Organic model kits or prepared models
        2   Chemistry handbooks

Laboratory Reagents and Supplies

1 Measurement and Significant Figures
    A. **Measuring Length**
        10   Lengths of string
    B. **Measuring Volume**
        3   Graduated cylinders partially filled with water and a few drops of food coloring
        10   Metal solids
    C. **Measuring Mass**
        10   Unknown mass samples

2 Density and Specific Gravity
    A. **Density of a Solid**
        10   Metal objects (cubes or cylinders of aluminum, iron, lead, tin, zinc, etc.)
        10   Large graduated cylinders to fit metal objects
        10   Strings or threads (short lengths to tie around metal solids)
    B. **Density of a Liquid**
        200 mL Isopropyl alcohol   200 mL Corn syrup
        200 mL 20% NaCl or 20% $CaCl_2$
        200 mL Mineral water, vegetable oil, milk, juices, soft drinks, window cleaners, etc.
    C. **Specific Gravity**
        3–4   Hydrometers in graduated cylinders containing water and the liquids used for (B).

3 Atomic Structure and Electron Arrangement
    A. **Physical Properties of Elements**
        Display of elements (metals and nonmetals)
    B. **Periodic Table**
        Periodic tables, colored pencils
        Display of elements

E. Flame Tests
- 10 Spot plates
- 10 Corks
- 10 Flame test wires
- 200 mL 1 M HCl
- 100 mL Unknown solutions (2–3) in dropper bottles (same as test solutions)

Place in dropper bottles:
- 100 mL 0.1 M CaCl$_2$
- 100 mL 0.1 M BaCl$_2$
- 100 mL 0.1 M CuCl$_2$
- 100 mL 0.1 M KCl
- 100 mL 0.1 M SrCl$_2$
- 100 mL 0.1 M NaCl

4 Nuclear Radiation

A. Background Count
- 1 Geiger-Müeller radiation detection tube

B. Radiation from Radioactive Sources
- 1 Geiger-Müeller radiation detection tube
- 3–4 Radioactive sources of alpha- and beta-radiation

Consumer products: Fiestaware®, minerals, old lantern mantles containing thorium compounds, camera lenses, old watches with radium-painted numbers on dials, fertilizers with P$_2$O, smoke detectors containing Am-241, anti-static devices for records and film containing Po-210, some foods such as salt substitute (KCl), cream of tartar, instant tea, instant coffee, dried seaweed

5 Chemical Reactions and Equations

A. Magnesium and Oxygen
- 10 Magnesium ribbon (2–3 cm)

B. Zinc and Copper(II) Sulfate
- 20 Pieces of zinc(s) strips, 1 cm square
- 100 mL 1 M CuSO$_4$

C. Metals and HCl
- 10 Cu pieces and Zn pieces 1 cm square
- 10 Mg ribbon pieces ~ 2 cm long
- 250 mL 1 M HCl

D. Reactions of Ionic Compounds
Place 100 mL each in dropper bottles:
- 0.1 M CaCl$_2$
- 0.1 M BaCl$_2$
- 0.1 M FeCl$_3$
- 0.1 M Na$_3$PO$_4$
- 0.1 M Na$_2$SO$_4$
- 0.1 M KSCN

E. Sodium Carbonate and HCl
- 20 Wood splints
- 25 g Na$_2$CO$_3$(s)
- 250 mL 1 M HCl

6 Moles and Chemical Formulas

A. Finding the Simplest Formula
- 10 Heat resistant pads
- 10 Mg ribbon (0.2–0.3 g, 16–18 cm strips)
- Steel wool

B. Formula of a Hydrate
- 10 Heat resistant pads
- 100 g Hydrate of MgSO$_4 \cdot$7H$_2$O

Appendix

7 **Energy and Matter**
   A. A Heating Curve for Water
      5   Timers
   B. Energy in Changes of State
      10   Styrofoam cups, covers
   C. Food Calories
      3–4   Food products with nutrition data on labels
   D. Exothermic and Endothermic Reactions
      50 g $NH_4NO_3(s)$ 50 g $CaCl_2(s)$ anhydrous
      10   Styrofoam cups, covers   10   Weighing papers
      10   Spatulas

8 **Gas Laws**
   B. Charles' Law
   Stoppers may be prepared by stock room with glass inserted attach to a short piece of rubber tubing
      10   one-hole stoppers to fit 125-mL Erlenmeyer flasks
      10   short pieces of glass   10   short pieces of rubber tubing
      10   pinch clamps   10   water pans
      20   boiling chips

9 **Partial Pressures of Oxygen, Nitrogen, and Carbon Dioxide**
   A. Partial Pressures of Oxygen and Nitrogen in Air
      25 g   Fe (iron) filings   1   Barometer
   B. Carbon Dioxide in the Atmosphere (*May be done as an instructor demonstration*)
   The following items may be assembled and available for use by instructor:
      10   Glass tubing (60–75 cm)
      10   Two-hole stoppers with two short pieces of glass tubing
      20   Rubber tubing (1 long, 1 short)   10 mL food coloring (optional)
      20 mL Mineral oil   100 mL 6 M NaOH

10 **Solutions, Electrolytes, and Concentration**
   A. Polarity of Solutes and Solvents (*May be a demonstration*)
      20 g $KMnO_4(s)$ 20 g $I_2(s)$
      20 g   Sucrose(s)   200 mL Cyclohexane
      20 mL Vegetable oil (2 dropper bottles)
   B. Solubility of KNO_3 (*Work in Teams*)
      10   10-mL graduated cylinders   100 g $KNO_3(s)$
      10   Weighing papers   10   Spatulas
   C. Concentration of a Sodium Chloride Solution
      10   10-mL Graduated cylinders (or pipettes)
      200 mL NaCl solution (saturated)

11 **Testing for Cations and Anions**
   A. Tests for Positive Ions (Cations)
      10   Spot plates   10   Flame test wires
      200 mL 3 M HCl   200 mL 6 M HNO_3
      100 mL 6 M NaOH
   Place 100 mL each in dropper bottles:
      0.1 M NaCl   0.1 M KCl
      0.1 M $CaCl_2$ 0.1 M $(NH_4)_2C_2O_4$

Materials and Solutions

0.1 M NH$_4$Cl	0.1 M FeCl$_3$	0.1 M KSCN

B. Tests for Negative Ions (Anions)
Place 100 mL each in dropper bottles:

0.1 M NaCl	0.1 M AgNO$_3$
6 M HNO$_3$	3 M HCl
0.1 M Na$_2$SO$_4$	0.1 M BaCl$_2$
0.1 M Na$_3$PO$_4$	0.1 M Na$_2$CO$_3$

(NH$_4$)$_2$MoO$_4$ reagent
50 mL Unknowns: KCl, Na$_2$CO$_3$, (NH$_4$)$_2$SO$_4$, CaCl$_2$, K$_2$SO$_4$, (NH$_4$)$_3$PO$_4$, etc.

12 Solutions, Colloids, and Suspensions
A. Identification Tests
200 mL	1% starch	200 mL	10% glucose
200 mL	10% NaCl	500 mL	Benedict's reagent

Place the following in dropper bottles:
100 mL	0.1 M AgNO$_3$	100 mL	Iodine solution

B. Osmosis and Dialysis
10	20-cm dialysis bag (cellophane tubing)		
100 mL	10% NaCl	100 mL	10% glucose
100 mL	1% starch	100 mL	0.1 M AgNO$_3$
100 mL	Benedict's solution	100 mL	Iodine solution

C. Filtration
25 g	Powdered charcoal	200 mL	1% starch
100 mL	Iodine solution		

13 Acids and Bases
A. pH Color Using Red Cabbage Indicator
1	Red cabbage	200 mL	Buffers pH 1–13

B. Measuring pH
2–3	pH meters	2–3	Wash bottles
2–3	Boxes Kimwipes	200 mL	Cabbage indicator from part A
20 mL	Buffers (pH 4, pH 10)		

Samples to test for pH: bring from home or have in lab. *Examples:* shampoo, hair conditioner, mouthwash, antacids, detergents, fruit juice, vinegar, cleaners, aspirin

C. Acetic Acid in Vinegar
10	5-mL pipet and bulbs	200 mL	Vinegar (white)
1 L	0.1 M NaOH (standardized)	100 mL	Phenolphthalein indicator

14 Alcohols, Aldehydes and Ketones
A. Structures of Alcohols and Phenol
10 Organic model kits or prepared models

Use for B. and C.
50 mL	Ethanol	50 mL	*t*-butyl alcohol (2-methyl-2-propanol)
50 mL	Cyclohexanol	50 mL	2-propanol
50 mL	20% phenol	2–4	Unknowns (Use same compounds)

C. Oxidation of Alcohols
100 mL 2% chromate solution

D. Properties of Aldehydes and Ketones
10	Organic model kits or prepared models		
50 mL	Acetone	50 mL	Benzaldehyde
50 g	Camphor	50 mL	Cinnamaldehyde,

Appendix

50 mL	Vanillin	50 mL	Propionaldehyde (propanal)
50 mL	Cyclohexanone	50 mL	2,3-Butanedione
2	Chemistry handbooks	2–3	Unknowns (use above compounds)

E. Iodoform Test for Methyl Ketones (Test tubes from part B.3)
Use compounds from B
100 mL 10% NaOH   200 mL Iodine test reagent

F. Oxidation of Aldehydes and Ketones
Use compounds from B
500 mL Benedict's reagent

16 Carbohydrates
A. Types of Monosaccharide's and B. Disaccharides
10   Organic model kits or prepared models

D. Benedict's Test for Reducing Sugars
200 mL 2% starch   500 mL Benedict's reagent
Place 50 mL each in dropper bottles:
2% glucose   2% fructose   2% sucrose   2% lactose
50 mL Unknown solutions from above *Examples*: glucose, fructose, sucrose, and lactose

E. Seliwanoff's Test for Ketoses
200 mL 2% starch   100 mL Seliwanoff's reagent
50 mL Unknown solutions from above *Examples*: glucose, fructose, sucrose, and lactose
Place 50 mL each in dropper bottles:
2% glucose   2% fructose   2% sucrose   2% lactose
50 mL Unknown solutions from above *Examples*: glucose, fructose, sucrose, and lactose

F. Fermentation Test
6   Fermentation tubes (or 6 small test tubes and 6 large test tubes),
200 mL 2% starch   40 g   Baker's yeast (fresh)
50 mL Unknown solutions from above
Examples: glucose, fructose, sucrose, and lactose
Place 50 mL each in dropper bottles:
2% glucose   2% fructose   2% sucrose   2% lactose
50 mL Unknown solutions from above *Examples*: glucose, fructose, sucrose, and lactose

G. Iodine Test for Polysaccharides
10   Spot plates
200 mL 2% starch   100 mL iodine reagent
Place 50 mL each in dropper bottles:
2% glucose   2% fructose   2% sucrose   2% lactose
50 mL Unknown solutions from above *Examples*: glucose, fructose, sucrose, and lactose

H. Hydrolysis of Disaccharides and Polysaccharides
10   Spot plate (or watch glasses)
100 mL 10% NaOH   100 mL 10% HCl
100 mL Iodine reagent   500 mL Benedict's reagent
Place 50 mL each in dropper bottles:
2% glucose   2% fructose   2% sucrose   2% lactose

17 Lipids
A. Triacylglycerols
5   Organic model kits or prepared models

B. Physical Properties of Some Lipids and Fatty Acids
25 g   Lecithin   25 g   Stearic acid
25 g   Cholesterol   100 mL Methylene chloride, CH_2Cl_2
Place 25 mL each in dropper bottles:

Oleic acid Vitamin A
Olive oil Safflower oil

C. Bromine Test for Unsaturation
5 Organic model kits or models Samples from B
100 mL 1% Br_2 in CH_2Cl_2

D. Preparation of Hand Lotion
Team project: Steps D.1, D.2, and D.3 may be prepared by different teams in the lab.
10 10-mL graduated cylinders 20 50-mL or 100-mL beakers
50 g stearic acid 15 g cetyl alcohol
25 g lanolin (anhydrous) 15 mL (dropper) triethanolamine
25 mL glycerin 100 mL ethanol
100 mL distilled water fragrance (optional)
commercial hand lotion products,

18 Saponification and Soaps

A. Saponification
Optional: Hot plate and a stirring bar Methyl Salicylate in a dropper bottle
100 mL 10% NaOH 100 mL 10% HCl
Blue litmus paper

B. Saponification: Preparation of Soap
Optional: Hot plate and a stirring bar 200 mL Ethanol
100 mL 20% NaOH 200 mL Saturated NaCl solution
10 pairs of disposable gloves
100 g Solid fats: lard, coconut oil, solid shortening, coconut oil
100 mL Liquid vegetable oils, olive or other vegetable oil

C. Properties of Soaps and Detergents
50 g Commercial soaps, lab-prepared soap (from part A), detergent
50 mL Safflower oil 100 mL 1% $CaCl_2$
100 mL 1% $MgCl_2$ 100 mL 1% $FeCl_3$

19 Amines and Amino Acids

A. Solubility of Amines in Water
Place in dropper bottles:
30 mL Aniline 30 mL Triethylamine
30 mL *N*-Methylaniline 100 mL 10% HCl

B. Neutralization of Amines with Acids
Test tubes from part B 1 100 mL 10% HCl

C. Amino Acids
10 Organic model kits or prepared models

D. Chromatography of Amino Acids
1 box Plastic wrap 1 box Whatman #1 filter paper (12 cm 5 24 cm)
50 Toothpicks or capillary tubing 1 Drying oven (80°C)
2 Hair dryers (optional) 1 Stapler
Place 50 mL each in dropper bottles:
1% Alanine 1% Glutamic acid
1% Serine 1% Aspartic acid
1% Lysine 1% Phenylalanine
50 mL 1% Unknown amino acids (Use samples from above list)
Chromatography solvent
100 mL 0.5 M NH_4OH 200 mL isopropyl alcohol
0.2% Ninhydrin spray reagent (in ethanol or acetone)

Appendix

20 Peptide and Proteins
A. Peptide Bonds and B. Structure of Proteins
    5   Organic model set
C. Denaturation of Proteins
Place in dropper bottle:

100 mL 1% egg albumin	Dissolve 1 g egg albumin in water to make 100 mL or students can make a fresh egg albumin solution by mixing the egg white from one egg with 200 mL of water and filtering the mixture through cheesecloth into a beaker.

100 mL 10% HNO_3 100 mL 10% NaOH
100 mL 95% ethanol   100 mL 1% $AgNO_3$

D. Isolation of Casein (Milk Protein)
200 mL Nonfat milk   200 mL 10% acetic acid

E. Color Tests for Proteins
10 g   Casein from part D
Place 100 ml each in dropper bottles:

    1% Glycine   1% Tyrosine
    1% Gelatin   1% Egg albumin (See part C.)
10% NaOH   50 mL   Conc. HNO_3
1 can   0.2% Ninhydrin reagent   200 mL 5% $CuSO_4$

21 Enzymes
A. Effect of Enzyme Concentration
    10   Spot plate (or plastic sheets)
    10   Timer   1% starch (buffered to pH 7)
    200 mL Amylase preparation   100 mL 1% Glucose
    100 mL Iodine reagent   500 mL Benedict's reagent

B. Effect of Temperature
Amylase preparation   200 mL 1% Starch
100 mL Iodine test reagent

C. Effect of pH
Amylase preparation   100 mL buffers (pH 2, 4, 7, 10)
200 mL 1% Starch   100 mL Iodine test reagent

D. Inhibition of Enzyme Activity
Amylase preparation
Place 50 mL each in dropper bottles:

    1% NaCl   1% $AgNO_3$
    1% $CuSO_4$ 1% $Pb(NO_3)_2$
    1% $HgCl_2$ Ethanol
200 mL 1% Starch   100 mL Iodine reagent

22 DNA Components and Extraction
A. Components of DNA
        10   organic model kits
B. Extraction of DNA
Different student teams may extract DNA from different DNA sources and compare results or the procedure may be run initially with onions and then repeated using other DNA sources selected by each student team.
DNA sources: white onions, cauliflower, broccoli, garlic, split peas, bananas, and/or animal sources such as chicken liver, calf thymus

Materials and Solutions

    5   knives   2   blenders
        cheesecloth or filter paper,   1-2   microscopes   ice
    1 L   SDS-NaCl-EDTA buffer solution (50 g sodium dodecyl sulfate, 50 g NaCl, 5 g sodium citrate, 0.2 mL of 0.5 EDTA and water to make 1 L)
500 mL isopropanol (in an ice bath)

Preparation of Solutions Used in the Laboratory

Acids and bases

Acetic acid $C_2H_3O_2$ (HAc)
       0.1 M HAc   Dilute 0.6 mL of glacial acetic acid with water to make 100 mL
       1 M HAc   Dilute 6 mL of glacial acetic acid with water to make 100 mL
       10% HAc   Dilute 50 mL of glacial HAc with water to make 500 mL

Ammonium hydroxide NH_4OH
       0.1 M NH_4OH Dilute 6.7 mL conc. NH_4OH with water to make 1.0 L
       0.5 M NH_4OH Dilute 34 mL conc. NH_4OH with water to make 1.0 L
       1 M NH_4OH Dilute 67 mL conc. NH_4OH with water to make 1.0 L

Hydrochloric acid HCl
       0.1 M HCl   Dilute 8.3 mL of conc. HCl with water to 1 L and standardize against
                        standardized 0.1 M NaOH (or 0.2 M NaOH)
       1.0 HCl   Dilute 85 mL conc. HCl with water to make 1.0 L
       2.0 M HCl   Dilute 170 mL conc. HCl with water to make 1.0 L
       3.0 M HCl   Dilute 250 mL conc. HCl with water to make 1.0 L
       10% HCl   Dilute 230 mL con. HCl to make 1.0 L

Nitric acid HNO_3
       6 M HNO_3 Dilute 76 mL conc. HNO_3 with water to make 200 mL
       10% HNO_3 Dilute 10 mL conc. HNO_3 with water to make 100 mL

Sodium hydroxide NaOH
       0.1 M NaOH   Dissolve 4.0 g NaOH in water to make 1.0 L
                        Standardization: Weigh a 1-g sample of potassium hydrogen phthalate, $KC_8H_5O_4$, to 0.001 g. Dissolve in 25 mL of water, add phenolphthalein indicator, and titrate with the prepared NaOH solution. Calculate the molarity (3 significant figures) as

$$\text{g phthalate} \; 5 \; \frac{1 \text{ mole phthalate}}{204 \text{ g phthalate}} \; 5 \; \frac{1}{\text{L NaOH used}} \; = \; \underline{\quad} M$$

       6 M NaOH   Dissolve 240 g NaOH in water to make 1.0 L
       3 M NaOH   Dissolve 120 g NaOH in water to make 1.0 L
       10% NaOH   Dissolve 10 g NaOH in water to make 100 mL
       20% NaOH   Dissolve 20 NaOH in water to make 100 mL

Salt solutions

Aluminum sulfate   1% $Al_2(SO_4)_3$ Dissolve 1.5 g $Al_2(SO_4)_3 \cdot 9H_2O$ in water to make 100 mL

Ammonium chloride   0.1 M NH_4Cl Dissolve 0.54 g NH_4Cl in water to make 100 mL

Appendix

Ammonium molybdate   Dissolve 8.1 g H_2MoO_4 in 20 mL water. Add 6 mL conc. NH_4OH to give a saturated solution. Filter. Slowly add filtrate to a mixture of 27 mL conc. HNO_3 and 40 mL water. Let stand 1 day. Filter and add water to 100 mL.

Reagent	Concentration	Preparation
Ammonium oxalate	0.1 M $(NH_4)_2C_2O_4$	Dissolve 1.4 g $(NH_4)_2C_2O_4 \cdot H_2O$ with water to 100 mL
Barium chloride	0.1 M $BaCl_2$	Dissolve 2.4 g $BaCl_2 \cdot 2H_2O$ in water to make 100 mL
Barium nitrate	0.1 M $Ba(NO_3)_2$	Dissolve 2.6 g $Ba(NO_3)_2$ in water to make 100 mL
Calcium chloride	1% $CaCl_2$	Dissolve 1.3 g $CaCl_2 \cdot 2H_2O$ in water to make 100 mL
	20% $CaCl_2$	Dissolve 200 g $CaCl_2$ in water to make 1.0 L
	0.1 M $CaCl_2$	Dissolve 1.5 g $CaCl_2 \cdot 2H_2O$ in water to make 100 mL
Copper(II) chloride	0.1 M $CuCl_2$	Dissolve 1.7 g $CuCl_2 \cdot 2H_2O$ in water to make 100 mL
Copper(II) sulfate	0.1 M $CuSO_4$	Dissolve 2.5 g $CuSO_4 \cdot 5H_2O$ in water to make 100 mL
	1 M $CuSO_4$	Dissolve 25 g $CuSO_4 \cdot 5H_2O$ in water to make 100 mL
	1% $CuSO_4$	Dissolve 0.78 g $CuSO_4 \cdot 5H_2O$ in water to make 50 mL
	5% $CuSO_4$	Dissolve 15.6 g $CuSO_4 \cdot 5H_2O$ in water to give 200 mL
Iron(III) chloride	0.1 M $FeCl_3$	Dissolve 2.7 g $FeCl_3 \cdot 6H_2O$ in water to make 100 mL
	1% $FeCl_3$	Dissolve 1.7 g $FeCl_3 \cdot 6H_2O$ in water to give 100 mL
Iron(III) nitrate	0.01 M $Fe(NO_3)_3$	Dissolve 0.40 g $Fe(NO_3)_3 \cdot 9H_2O$ in water to make 100 mL
	1 M $Fe(NO_3)_3$	Dissolve 40. g $Fe(NO_3)_3 \cdot 9H_2O$ in water to make 100 mL
Lead(II) nitrate)	1% $Pb(NO_3)_2$	Dissolve 0.5 g $Pb(NO_3)_2$ in water to make 50 mL
Magnesium chloride	1% $MgCl_2$	Dissolve 2.1 g $MgCl_2 \cdot 6H_2O$ in water to make 100 mL
Mercury(II) chloride	1% $HgCl_2$	Dissolve 0.5 g $HgCl_2$ in water to make 100 mL
Potassium chloride	0.1 M KCl	Dissolve 0.75 g KCl in water to make 100 mL
Potassium permanganate	1% $KMnO_4$	Dissolve 1 g $KMnO_4$ in water to give 100 mL
Potassium thiocyanate	0.01 M KSCN	Dissolve 0.097 g KSCN in water to make 100 mL
	0.1 M KSCN	Dissolve 0.97 g KSCN in water to make 100 mL
	1 M KSCN	Dissolve 9.7 g KSCN in water to make 100 mL
Silver nitrate	0.1 M $AgNO_3$	Dissolve 1.7 g $AgNO_3$ in water to make 100 mL
	1% $AgNO_3$	Dissolve 0.5 g $AgNO_3$ in water to make 50 mL
Sodium carbonate	0.1 M Na_2CO_3	Dissolve 2.9 g $Na_2CO_3 \cdot 7H_2O$ in water to make 100 mL
Sodium chloride	0.1 M NaCl	Dissolve 0.58 g NaCl in water to make 100 mL
	1% NaCl	Dissolve 1 g NaCl in water to make 100 mL
	10% NaCl	Dissolve 10 g NaCl in water to make 100 mL
	20% NaCl	Dissolve 200 g NaCl in water to make 1.0 L
	Saturated NaCl	Add 80 g NaCl to water to make 200 mL
Sodium phosphate	0.1 M Na_3PO_4	Dissolve 3.8 g $Na_3PO_4 \cdot 12H_2O$ in water to make 100 mL
Sodium sulfate	0.1 M Na_2SO_4	Dissolve 3.2 g $Na_2SO_4 \cdot 10H_2O$ in water to make 100 mL
	1% Na_2SO_4	Dissolve 2.3 g $Na_2SO_4 \cdot 10H_2O$ in water to make 100 mL
Strontium chloride	0.1 M $SrCl_2$	Dissolve 1.95 g $SrCl_2 \cdot 2H_2O$ in water to make 100 mL

Carbohydrates

Fructose	2% fructose	Add 1 g fructose to water to make 50 mL
Glucose	0.1 M glucose	Dissolve 1.8 g glucose in water to make 100 mL
	2% glucose	Add 1 g glucose to water to make 50 mL
	10% glucose	Dissolve 10 g glucose in water to make 100 mL
Lactose	2% lactose	Add 1 g lactose to water to make 50 mL
Sucrose	0.1 M sucrose	Dissolve 3.42 g sucrose in water to make 100 mL
	2% sucrose	Add 1 g sucrose to water to make 50 mL
Starch	1%	Make a paste of 2 g soluble starch and 40 mL water. Add to 160 mL of boiling water to make 200 mL. Stir and cool.
	2%	Make a paste of 4 g soluble starch and 40 mL water. Add to 160 mL of boiling water to make 200 mL. Stir and cool.

Reagents

Benedict's reagent — Dissolve 86 g sodium citrate, $Na_3C_6H_5O_7$, and 50 g anhydrous Na_2CO_3 in 400 mL water. Warm. Dissolve 8.6 g $CuSO_4 \cdot 5H_2O$ in 50 mL water. Add to sodium citrate solution, stir, and add water to make 500 mL solution.

DNA buffer — SDS-NaCl-EDTA buffer solution (50 g sodium dodecyl sulfate, 50 g NaCl, 5 g sodium citrate, 0.2 mL of 0.5 EDTA and water to make 1 L

0.15% Salicylic acid — Dissolve 0.15 g salicylic acid in 100 mL water.

Seliwanoff's reagent — Dissolve 0.15 g resorcinol in 100 mL 6 M HCl

Iodine solution — Dissolve 10 g I_2 + 20 g KI in water to make 500 mL

Iodoform (iodine solution) Dissolve 10 g I_2 + 20 g KI in water to make 100 mL

0.2% Ninhydrin — Dissolve 0.2 g in ethanol to make 100 mL

20% Phenol — Dissolve 20 g phenol in water to make 100 mL

4% Hexamethylenediamine/NaOH
Combine 3.0 g $H_2N(CH_2)_6NH_2$ + 1.0 g NaOH and add water to 50 mL

4% Sebacoyl chloride/Hexane
Combine 20.0 mL $ClCO(CH_2)_8COCl$ and hexane to make 50 mL

Indicators

1% Phenolphthalein — Dissolve 1 g phenolphthalein in 50 mL ethanol and 50 mL water

1% Bromine solution — Dilute 1 mL Br_2 with methylene chloride to make 100 mL

2% Chromate — Dissolve 2.0 g $K_2Cr_2O_7$ in 10 mL of 6 M H_2SO_4; then carefully add to water to make 100 mL

Amino acids

1% Alanine — Dissolve 0.5 g alanine in water to make 50 mL
1% Aspartic acid — Dissolve 0.5 g aspartic acid in water to make 50 mL
1% Glutamic acid — Dissolve 0.5 g glutamic acid in water to make 50 mL
1% Glycine — Dissolve 1 g glycine in water to make 100 mL
1% Lysine — Dissolve 0.5 g lysine in water to make 50 mL
1% Phenylalanine — Dissolve 0.5 g phenylalanine in water to make 50 mL
1% Serine — Dissolve 0.5 g serine in water to make 50 mL

Appendix

1% Tyrosine   Dissolve 1 g tyrosine in water to make 100 mL

Proteins

1% Egg albumin   Dissolve 1 g egg albumin in water to make 100 mL
1% Gelatin   Dissolve 1 g gelatin in water to make 100 mL
2% Pancreatin   Dissolve 2.0 pancreatin in 100 mL 0.25% Na_2CO_3. Use immediately.
2% Pepsin   Dissolve 2% pepsin in water to make 100 mL